SACRED SYMBOLS OF THE DOGON

SACRED SYMBOLS OF THE DOGON

THE KEY TO ADVANCED SCIENCE IN THE ANCIENT EGYPTIAN HIEROGLYPHS

LAIRD SCRANTON

Foreword by John Anthony West

Inner Traditions
Rochester, Vermont

Inner Traditions
One Park Street
Rochester, Vermont 05767
www.InnerTraditions.com

LIBRARY OF CONGRESS CATALOGING-IN-PUBLICATION DATA

Scranton, Laird, 1953–
Sacred symbols of the Dogon : the key to advanced science in the ancient Egyptian hieroglyphs / Laird Scranton.
p. cm.
Summary: "Dogon cosmology provides a new Rosetta stone for reinterpreting Egyptian hieroglyphs"—Provided by publisher.
Includes bibliographical references and index.
ISBN-13: 978-1-59477-134-7 (pbk.)
ISBN-10: 1-59477-134-0 (pbk.)
1. Egyptian language—Writing, Hieroglyphic. 2. Mythology, Dogon. 3. Science—Egypt—History—To 1500. I. Title.
PJ1097.S37 2007
493'.1—dc22

2007021384

Printed and bound in the United States by Lake Book Manufacturing

10 9 8 7 6 5 4 3 2 1

Text design and layout by Virginia Scott Bowman
This book was typeset in Sabon with Schneidler Initials and Agenda as display typefaces

To send correspondence to the author of this book, mail a first-class letter to the author c/o Inner Traditions • Bear & Company, One Park Street, Rochester, VT 05767, and we will forward the communication.

CONTENTS

FOREWORD

The history of scholarship and science is punctuated by examples of the rank amateur who, fired by curiosity, decides to explore some aspect of a discipline that is not his own but that has not been resolved or, in some cases, even noticed.

In successful stories (Schliemann's discovery of ancient Troy may be the best known), curiosity plus persistence produces initial results. The researcher is inspired to press forward, mastering the disciplines necessary to continue the work, which eventually leads to something totally unexpected, compelling, and significant.

Laird Scranton's quest followed this documented but by no means well-traveled path.

Initially, intrigued by a suggestion that the intricate, coherent (still living and still practiced) cosmology of the remote West African Dogon tribe was in some way related to that of ancient Egypt (four thousand miles across the North African Sahara and with no demonstrable past or present connection to the Dogon), he began his research.

This notion was in no way directly related to his day job. Scranton is an expert in computer software design. When a company wants to make customized changes to their software, Scranton interprets their existing system of codes and files and finds a way to make the changes. So, in retrospect, he was actually the perfect curious amateur to take on the task—uncommitted to a particular point of view but with the specialized symbolic skills, in principle, to carry it out.

Relationships and equivalents between Dogon and Egyptian traditions were quickly apparent to Scranton. Not only were their mythological accounts similar, but even the words and symbols used to describe the sequential stages of cosmogenesis (the birth of the cosmos) were nearly the same. In other words, the contemporary tribal Dogon and the ancient, highly organized, and accomplished ancient Egyptians had the same belief system. (Scranton's research also turned up evidence of that same system in several other traditions as well.)

In and of itself, this was a revolutionary conclusion that was carefully documented, logically developed, and compelling. It was at this point, sparked again both by curiosity and by the explicit Dogon assertion that their myths described the formation and structure of matter, that Scranton decided to look into contemporary, cutting-edge cosmological science to see how that compared with these ancient mythologies.

To his astonishment, although the languages used by modern and ancient cosmologists were very different (mathematics for the moderns, myth and symbol for the ancients), the stories and sequences were effectively identical; in other words, the ancients were talking about relativity, quantum mechanics, and even currently controversial string and/or torsion theory in their own languages of myth and symbol. It was all there: the big bang (or an equivalent thereof), waves that become particles, subatomic particles, the electron, the atom . . . all of it. This was the theme of *The Science of the Dogon.*

However, because this amounted to a direct rebuttal of the central dogma of Western civilization—progress as a linear process going from primitive beginnings to ourselves—it was predictably ignored by mainstream science and the mainstream media.

The original self-published edition *Hidden Meanings: A Study of the Founding Symbols of Civilization* reached a small but informed audience, which widened considerably after it was acquired by Inner Traditions International and republished as *The Science of the Dogon.* Correspondents provided cosmological similarities from the mythologies of other societies, ancient and contemporary, and suggested new sources for Scranton to explore. The most striking of these was a study of the symbolism of the traditional Buddhist *stupa,* the conical structure very similar in shape, symbolism, and function to the earthen Dogon granary (which tells the Dogon cosmological story in symbolic form). The Buddhist version is effectively the same and was studied in detail by Adrian Snodgrass, an Australian architect and Buddhism scholar, in his book *The Symbolism of the Stupa.* Although Snodgrass did not explicitly link the traditional and contemporary accounts, he affirmed that the ancient cosmogenesis story was meant to symbolize and "mimic" the processes of the formation of matter, an interesting choice of words because, short of postulating precognitive mimicry, it is possible to mimic only that which is already there. The plain fact is that traditional Buddhist cosmology (originating no one knows when but long before our astrophysics arrived on the scene) also tells the same story that contemporary astrophysicists are only now in the process of developing.

The evidence is now commanding: the Einsteins of old, whoever they may have been and whenever they lived, had a cosmological science as advanced and as exact as our own and that knowledge, at some point long ago, was global in reach.

This barrage of additional backup was gratifying, of course, but in a sense superfluous—like those first photos from space showing that the Earth was, after all, round. The evidence had been put together well before the shuttle went up, but even so it was exciting to see it confirmed with our own eyes. So it was with the claim that advanced civilizations existed in the very distant past. There was ample evidence already, but Scranton's work is the equivalent of that space shot.

Scranton was, however, more interested in going deeper rather than wider, and that is the focus of *Sacred Symbols of the Dogon.* The Egyptian hieroglyphic system is the only remaining wholly ideographic language with (mostly) recognizable symbols (bird, animal, human, plant, tool, etc.) individually or together representing words that express concepts, ideas, actions, and material objects. Ironically, hieroglyphics had disappeared entirely as a written and spoken language, whereas other languages that were probably similarly ideographic originally, but no longer are (Chinese and Sanskrit, for example), still exist as languages. An ancient text written in one of those languages can be easily deciphered by a modern scholar, although the ideographic original has been lost. The hieroglyphs, on the other hand, are all still there, perfectly intact, but their meanings had to be rediscovered from scratch, mainly by scholars convinced that "real" civilization began with the Greeks and, in any case, without access to (or much interest in) the latest developments in physics and cosmology.

Scranton, however, recognizing the advanced science underlying both Egyptian and Dogon mythology and symbolism, wondered if the entire system might be grounded in cosmogenesis, extending by logical steps—in a system of cascading analogies and metaphors—from astrophysics to terrestrial physics to the activities and manifestations of daily life. And, in fact, so they do.

It suddenly becomes clear why specific glyphs were carefully chosen for specific roles. (Scranton uses standard accepted translations of words and glyphs in building his case; he is not retranslating anything to make the conclusion fit his hypothesis.) It becomes equally clear why certain combinations of these glyphs were chosen to produce specific words. There was nothing arbitrary about the process.

What before seemed alien, impenetrable, and remote from the concerns of daily modern life suddenly becomes lucid, meaningful, and, in the true sense of that word, enlightening. It all suddenly makes sense when the hieroglyphs reveal their deepest inner secrets and act as intermediaries between the very new and the very old.

Those who follow the argument (easy enough to understand but requiring persistence to stay the course) will be treated to a rare work of scholarly detective nonfiction: Sherlock Holmes meets Champollion 2.0.

PS: As an unexpected, added bonus, persistent nonspecialists and the advanced-science-challenged (such as myself) will also end up with a far better grasp of these arcane modern disciplines than they would even from those works specifically designed for lay readers and written by the scientists themselves.

John Anthony West

John Anthony West is an independent Egyptologist who has studied and written about ancient Egypt since 1986. His controversial work on redating the Great Sphinx has challenged long-accepted notions of Egyptian history. He is the author of *Serpent in the Sky* and *The Traveler's Key to Ancient Egypt* and often leads tours to Egypt as a guide and lecturer.

ACKNOWLEDGMENTS

This book would not have been possible without the support, patience, and understanding of others. In particular, I greatly appreciate the support of my wife Risa, son Isaac, and daughter Hannah, who have served as sounding boards for my ideas throughout many years of research. I also owe a great debt of thanks to John Anthony West, whose steadfast efforts—both on my behalf and on behalf of the larger community of researchers and writers of which I am a part—are greatly appreciated. I would like to thank Edmund Meltzer, who served as a paid consultant on this project and whose many personal and professional insights have benefited this work enormously. I am indebted to John Dering for his willingness to review this text with an eye toward theories of physics and for his generous insights about those theories and how they might relate to ancient myths. I am grateful to Joscelyn Godwin at Colgate University for his willingness to share my ideas with his insightful students. I would also like to express my gratitude in general to the many friends, relatives, and acquaintances who have shown enthusiasm and support for my studies, including my brother David Scranton, Will Newman and Sue Clark, Bill Churchman, Dave Carl, John Gardiniere, Eric Infante, Jim Valenti, Shawn Francis, Teresa Vergani, Ida Moffett Harrison, Mr. Nataki at Afrikan World Books in

Baltimore, William Henry, Elizabeth Newton, and Alan Glassman. Last but not least, I would like to dedicate this book to former teachers and mentors Don Ball and Lee Wells, who many years ago set me on the path of intellectual inquiry that led to these studies.

ONE

INTRODUCTION TO DOGON SCIENCE IN THE EGYPTIAN HIEROGLYPHS

This book, *Sacred Symbols of the Dogon*, is the second in a series—it follows a 2006 volume titled *The Science of the Dogon: Decoding the African Mystery Tradition,* which examines the cosmology of a modern-day African tribe called the Dogon. Although *The Science of the Dogon* focuses primarily on the tribal myths and cosmological symbols of the Dogon, it also documents many consistent resemblances between Dogon mythological keywords and symbols and those of the ancient Egyptian hieroglyphic language. Examples presented in *The Science of the Dogon* (and summarized in the early chapters of this book) demonstrate the consistency of these resemblances and show they are rooted in key symbols shared commonly by both cultures. In some cases, likely identities can be established between Dogon and Egyptian words based on common pronunciation and meaning explicitly defined for each word in its respective language. *The Science of the Dogon* also demonstrates a consistent relationship between key drawings within Dogon mythology and the shapes of specific Egyptian glyphs. Although we might speculate indefinitely about the theoretic symbolism that might apply to an Egyptian glyph, the meanings of the corresponding mythological shapes are well known to the Dogon priests, their definitions clear and specific in the Dogon tradition. So apart from any broader discussion of Dogon tribal myths, the suggestion is that these definitions, which have been well preserved in Dogon myth and language, might provide an alternate point of entry for the interpretation of Egyptian hieroglyphs. This premise of a relationship between Dogon and Egyptian symbols and languages defines the central theme for this second volume—*Sacred Symbols of the Dogon*.

What first draws our attention to the Dogon of Mali are the many intimate details of a complex cosmology—extensively defined in myths, symbols, rituals, and drawings—that describe the efforts of a tribal god named Amma to create the universe and the matter it contains. This cosmology is part of a larger Dogon tradition whose defined purpose is to document the successive stages of a mythological process of creation. What had gone unnoticed prior to *The Science of the Dogon* were the many consistent resemblances between the descriptions of the Dogon myths and actual stages in the formation of matter as defined by modern-day science. *The Science of the Dogon* demonstrated in many different ways that the Dogon descriptions of these processes are scientifically accurate. They correctly define the key components of matter from atoms to quarks to the vibrating threads of string theory—all in the proper sequence, each with its appropriate scientific attributes and many supported by explicit tribal drawings that often replicate pertinent scientific diagrams.

Even from a more traditional viewpoint, Dogon mythology may have great relevance to studies of the ancients because it stands at the crossroads of several important traditions. The Dogon share many common elements of culture, mythology, and language with ancient Egypt, and they observe many of the same religious traditions and rituals as ancient Judaism. Likewise, the stories of Dogon mythology gravitate around a rich set of classic themes and symbols that are already familiar to us from other world mythologies and religions—the spiraling coils, clay pots, and primordial serpents that reappear again and again among the myths and stories of ancient cultures. Key elements of Dogon mythology bear great resemblance to those found in other cultures, including those of ancient Greece, Sumer, India, and China, some Polynesian cultures such as the Maori of New Zealand, and even the Olmec and Mayan cultures of the Americas.

Among all of these cultures, the Dogon are of particular interest because they survive to this day as a cohesive society—still seemingly observant of some of the most ancient traditions and yet also able to express those traditions in coherent, modern terms. Although we as modern researchers are not in a position to pose direct questions to an ancient Egyptian priest, we are quite able to converse intelligently with a modern-day Dogon priest about subjects that may relate directly to Egyptian culture and language. Surely it would behoove us to pay close attention to what the Dogon priests may have to say about symbols, rituals, myths, and language, because their insights could potentially open a new window for us on an ancient era of Egyptian mythology and culture that, in many other respects, has long since passed from view.

The many compelling aspects of Dogon life that recall those of ancient cultures suggest that we should view the Dogon as a kind of modern remnant of a very ancient tradition—almost as a living artifact of times long past. We should be clear that it is not the purpose of this study to trace a direct or indirect lineage between Dogon and Egyptian cultures or to endorse any larger psychological mechanism by which two unrelated cultures might adopt a similar set of symbols, words, or rituals. Rather, our primary focus should be on what is, not how it came to be. As a general rule, for the purposes of these discussions, if two different cultures assign the same mythological meanings to a similarly-shaped symbol, we will consider both references to be to the same symbol—no matter that the two cultures might be separated by half a world in distance or five thousand years in time. Likewise, if both cultures assign the same set of meanings to similarly pronounced words, we will consider the roots of these words to be common to both cultures. The presumption is that two truly unrelated cultures should not sustain consistent similarities of words, symbols, myths, and mythological meanings. When they do, some degree of relationship is reasonably inferred—whether it is one of common origin, borrowing, or diffusion by some other method. It is not necessary to prove that Dogon society is as old as ancient Egypt; rather, it is sufficient to observe that the rituals and traditions sustained by the Dogon are known to date from a period as early as 3000 BCE and that the Dogon define and discuss these traditions using words and symbols that bear a close resemblance to those from the same remote period of Egyptian antiquity.

Similarities between Dogon and Egyptian culture manifest themselves at nearly every level of comparative study. They can be found in the simple pronunciations and meanings of individual words, in the parallel acts and phonetically similar names of equivalent mythological gods, and in the nearly identical forms of many key mythological symbols. We often find direct correspondences between the ritual practices of the Dogon and Egyptian religions, their agricultural methods, and the shared elements of their ritual structures. Parallels between these two cultures extend even to the guiding principles around which their civil bodies were established—in deliberate pairs, one designated as Upper and the other as Lower. Such correspondences between the Dogon and Egyptian systems are so pervasive that, when in doubt, one might reasonably assume that if a given practice exists among the Dogon, it most likely also existed in similar form with the ancient Egyptians.

Although it is the similarity of Dogon and Egyptian culture that drives our initial comparisons of their myths and symbols, it is often the subtle differences between cultures that illuminate those comparisons. For example, the Egyptian hieroglyphic language may seem forbidding and obscure, especially when compared to the Dogon language, which employs far fewer words and is documented in terms of an alphabet rather than pictographic glyphs. Yet many important mythological concepts in each of these languages are formed from similar root words, pronounced in similar ways, and carry similar sets of meanings. So when we are faced with a complex or uncertain Egyptian word or concept, we find that we often can gain insight into its meaning by looking to corresponding words in the more accessible Dogon language, whose meanings have been clearly and overtly defined. A similar approach can be taken with the study of Dogon and Egyptian myths and symbols. If the surviving text of an Egyptian myth seems fragmented or obscure, we find that there may well be corresponding episodes in Dogon mythology that can provide us with clarifying details.

Like many ancient religions, the Dogon tradition includes both public and private aspects. The details of Dogon cosmology present themselves first through a body of exoteric myths (fireside stories known to most Dogon tribe members) that describe in a general way the efforts of the god Amma to create the sun, the Earth, the moon, and the spiraling galaxies of stars and planets. These story lines run parallel to a more detailed set of esoteric myths (those known primarily to the Dogon priests) that lay out the hierarchy of a complex cosmological system in an intricate system of symbols, signs, drawings, and keywords. The innermost details of this system are carefully sheltered from public view and are revealed only to potential initiates of the religion—candidates who have been carefully screened by the Dogon priests. Above all else, the salient quality sought in a potential initiate to the Dogon religion is that he or she demonstrates an abiding curiosity about the religion itself, a quality that is most often expressed by the persistent asking of questions. In truth, the Dogon priests are obliged by tradition to faithfully answer any orderly question posed by an initiate. Over time, this priestly obligation became the cornerstone of an instructional dynamic in which knowledge would be divulged to an initiate only after the candidate asked the appropriate question. In this way, for learning to progress between a student and a priest, it became the implied job of the student to ask the next question.

Much of what we know about Dogon mythology and culture comes from the studies of two French anthropologists, Marcel Griaule and Germaine Dieterlen, who lived and worked among the Dogon in Mali for many years, from the 1930s through the mid-1950s. Over the years, Griaule—in his role as an anthropologist—came to be regarded as a close, trusted friend of the Dogon tribe, and he was prone, both by profession and by nature, to ask persistent and penetrating questions. This inquisitive side of Griaule's personality qualified him in a real sense to become an initiate of the religion and eventually induced a blind Dogon priest named Ogotemmeli to seek the permission of a priestly council to instruct Griaule in the more closely guarded traditions of the religion. Griaule wrote the book *Dieu d'Eau* (whose English language title is *Conversations with Ogotemmeli*), a diary based on his thirty-three days of instruction by the Dogon priest. Griaule and Dieterlen's research also produced another, more substantial work about the Dogon religion, upon which both *The Science of the Dogon* and *Sacred Symbols of the Dogon* are based, called *Le Renard Pale* (in English, *The Pale Fox*). This finished anthropological study of the Dogon religion, compiled by Dieterlen after the death of Griaule in 1956, ultimately came to include contributions not only from the priests of the Dogon but also from those of three other closely related tribes, all of whom share similar cosmologies and traditions. Wherever appropriate, Griaule and Dieterlen chose to include alternate versions of myths and minority opinions about key passages within the myths, intending that the study reflect a true consensus of Dogon thought relating to the episodes of their cosmology.

The Pale Fox can be a difficult volume—not because of any fault in its clarity but because its structure may have been deliberately designed to reflect many of the deep enigmas of the Dogon religion itself. It is possible that at the time of its composition, Dieterlen found herself in an intractable dilemma—caught between the ethics of a professional anthropologist and the moral obligation of her colleague Griaule as a Dogon initiate. Professionally, she was required to report the many intimate details she had learned about the Dogon religion, but as a caretaker of privileged tribal knowledge she may also have felt morally bound to uphold its mysteries. Her thoughtful solution to that ethical quandary may have been to compose an anthropological study that was carefully designed to safeguard its own inner secrets and that, like the Dogon religion itself, offered answers only to the most persistent of questioners. And so it is within this context that we find several key Dogon drawings—some of which Dieterlen herself describes as fundamental to a correct understanding of the Dogon religion—distributed widely among diverse chapters of the study, often labeled in generic ways, and sometimes tucked away unceremoniously in the body of an appendix to the book.

A third text that has been essential to this study is Genevieve Calame-Griaule's dictionary of the Dogon language, *Dictionnaire Dogon*. This dictionary includes comprehensive entries that define each of the words of the Dogon *Toro* dialect. One discovers as they work with *Dictionnaire Dogon* that a key Dogon word often carries two levels of definition—one appropriate to its meaning in common usage, and one appropriate to its mythological meaning. These same definitions are often found as multiple word-entries in an Egyptian hieroglyphic dictionary, pronounced in a similar way. This attribute, common to both languages, allows us to draw likely equivalencies, and sometimes outright identities, between Egyptian words and their Dogon counterparts, based on consistencies of pronunciation and meaning.

Before we begin, there are three major aspects of this study that might elicit an immediate objection from the knowledgeable reader. The first is its reliance on Griaule and Dieterlen's Dogon cosmology as a primary source. They documented what they characterized as a secret Dogon cosmology, preserved as an esoteric tradition by the Dogon priests and shared with a few select Dogon initiates. According to Griaule and Dieterlen, this cosmology is founded on an aligned ritual structure called a granary. They reported that the details of this cosmology are largely unknown to the average Dogon tribesperson and are held as a closely guarded secret by the more informed Dogon priests. Later researchers who have worked among the Dogon state that they have been unable to confirm Griaule and Dieterlen's findings—this notwithstanding Griaule's original characterization of the cosmology as a secret tradition. In correspondence to me, Belgian anthropologist Walter Van Beek referred to the classic plan of the Dogon granary as "a chimera known only to Griaule," and rather than allowing the possibility that Griaule might have, in good faith, documented a legitimate but well-kept secret tradition to which Van Beek and other researchers might not be privy, concluded that Griaule or the Dogon priests must have fabricated the cosmology. In Professor Van Beek's view, if Griaule's cosmology constitutes a form that was invented wholesale by Griaule, it should be disallowed as the primary reference for a study such as this one.

However, there is direct evidence to support Griaule's secret Dogon tradition as the reflection of a wholly legitimate cosmological form that is either unknown to or has gone unreported by Van Beek and other researchers. This evidence is presented by Adrian Snodgrass of the University of West Sydney, Australia, who is a leading authority on Buddhist architecture and symbolism. In his book, *The Symbolism of the Stupa,* Professor Snodgrass describes a traditional, aligned ritual structure found commonly across India and Asia whose base plan matches that of Griaule's Dogon granary and that evokes a complex symbolic system that is a near-exact match for Griaule's Dogon cosmology. Like the esoteric Dogon tradition that Griaule describes, the stupa's symbolism is preserved primarily by the most knowledgeable Buddhists; like Griaule's cosmology it evokes parallel symbolic themes that are described as cosmological and biological in nature; like Griaule's cosmology, it defines matter as a product woven by a spider that evolves like rays of a star and is characterized as a spiraling coil. In other words, the Buddhist symbolism as defined by Snodgrass confirms virtually all key aspects of Griaule's Dogon cosmology.

The parallels between Snodgrass's and Griaule's cosmologies effectively preclude the possibility that either Griaule or the Dogon priests could have simply invented what Griaule reports as Dogon cosmology. Likewise, in the fifty years since Griaule's death, it would appear that not a single West African anthropologist has attained an awareness of Buddhist symbolism sufficient to recognize the near-exact match to Griaule's Dogon cosmology. I, myself, required fifteen years study of comparative mythologies before becoming aware of the match. Griaule could not have known Snodgrass's study, which was published more than twenty years after Griaule's death, and biographical sketches report no long period of Buddhist study for Griaule. So, in the absence of direct evidence that Griaule was specifically aware of the intricacies of the Buddhist symbolism, any suggestion that he could have fabricated his Dogon cosmology is specifically contradicted. It seems far more likely that Griaule faithfully reported a legitimate tradition that was honestly conveyed to him by his Dogon informants and that modern Dogon contacts have simply lied to later researchers to protect a longstanding, important tribal secret.

The next aspect of this study that might elicit an objection involves the use of an unproved—albeit promising—concept such as string theory as a basis of comparison with the symbols of Dogon cosmology. The justification for the use of string theory is perhaps best understood in terms of a brief analogy. Imagine that you are a high school student hosting a first-time visit from a five-year-old guest who confidently announces that he can open the combination lock to your locker. You realize that such a claim should be beyond the reasonable capability of a five-year-old, especially one who has never visited the high school before. Because there can be no harm in simply letting the child *try* to open the lock, you agree to let him do so. Imagine then, that the child steps up to the locker, turns the dial of the lock to the right the correct number of full turns until he stops at the first number of the combination, turns it left to the correct second number, turns it right again to the third correct number, and then actually opens the lock. All objections and protests aside, the child has just demonstrated positive knowledge of his subject. He successfully opened the lock, just as he predicted he would. So unless a person were willing to invoke the highly unlikely possibility of an extreme series of coincidences that somehow allowed the child to simply guess the numbers, there is really no disputing what the child has done. The proof is in the act, and clearly he knows what he says he knows.

When it comes to the suggestion of any real relationship between mythology and science, in our eyes Dogon mythology is that five-year-old child. From the very outset, our reasonable assumption is that the Dogon myths cannot possibly describe actual details of science. However, like the child in our analogy, the Dogon mythology clearly states its intention to describe how matter is formed, then proceeds to define component after component of matter, each in its proper sequence and each definition supported by appropriate descriptions, attributes, and drawings. Moreover, if we follow this hierarchy of components to its point of origin, we arrive at a fundamental component of matter whose description is in explicit agreement with string theory: a tiny vibrating thread that the Dogon claim to be the source of all matter. The idea that matter is woven from threads constitutes one of the classic themes of ancient mythology, one that is as old as Egyptian mythology itself. In the Egyptian mythological tradition, it is the great mother goddess Neith who is credited with having woven matter with her shuttle. What may be less obvious to students of mythology is that the inherent scientific nature of this theme is explicitly supported by the Egyptian language. As *The Science of the Dogon* shows, the hieroglyphic name of Neith and the related Egyptian words for "weaving" and "thread" are written with glyphs that clearly depict the three precise ways that the strings are believed to intersect each other in string theory.[1]

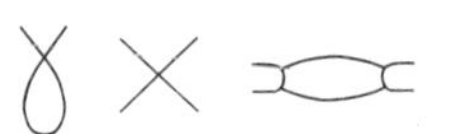

Three types of string intersections

The probability that these cultures might have just coincidentally guessed the key tenet of string theory—that all matter is a by-product of vibrating threads—goes well beyond that of a five-year-old child guessing the combination of a lock. However, for the sake of discussion, let us take the most generous view and concede that coincidences *do* happen and allow for the possibility of an extremely fortuitous guess. Could any thinking person then seriously suggest that these same cultures also managed to blindly guess the shapes of the three types of string intersection, diagram them correctly, and place them in the correct scientific context—the weaving of matter? Surely not. At some point, simple reason requires that we accept that the five-year-old might know what he says he knows—that these mythological definitions represent an overt expression of positive knowledge. If that were the case, then what struck us at first as simple tribal symbols would take on a much larger significance in relation to science because, in effect, they would constitute an independent verification of string theory. That, ironically, puts a reverse spin on the use of string theory in this study. In this case, there is the very real potential that the symbols of myth will ultimately validate specific details of string theory.

In truth, the cosmological system of the Dogon as explicated by the Egyptian language includes important elements that are found both in string theory and in earlier asymmetric field theories that included the concepts of torsion and distant parallelism, as proposed by Albert Einstein and others. More recently, these theories have been extended and adapted by researchers such as Jim Corum, Pharis Williams, and Gennady Shipov. Among these elements is the notion of coiled and twisted spaces, which are suggestive of the Calabi-Yau space of string theory and also of the physical relationships between gravitation, torsion, mechanical spin, electromagnetism, and rotating electromagnetic fields described in early unified field theories.

The third aspect of *Sacred Symbols of the Dogon* that might raise an objection from a traditional Egyptologist is the use of Sir E. A. Wallis

Budge's *An Egyptian Hieroglyphic Dictionary* as a primary reference for the Egyptian hieroglyphs. Budge's dictionary was compiled in the nineteenth century and is now considered for many reasons to be out of date and in some respects unreliable. Over the years and partly due to the ongoing discovery of many additional Egyptian texts, theories of the Egyptian language have moved well beyond Budge in several different respects, some of which involve the way he vocalizes various Egyptian glyphs. Many of the researchers who came after Budge rely more heavily on comparisons to Coptic, an Egyptian language that is closely related to the hieroglyphs and from which Egyptian scholars often draw linguistic parallels.

I should emphasize that my use of Budge's text is primarily as a tool for aligning Dogon cosmological symbols and keywords with those used in ancient Egypt. One feature of Budge's dictionary that is indispensable to this process is the tendency of his word entries to accurately reflect African pronunciations and meanings. A good case in point is found with a key African word that is integral to the cosmology—*Amma* (or *Amen*). The Dogon define two primary meanings for this word. It is the name of their "hidden god" who is regarded as too small to see, and it is also defined as meaning "to grasp, to hold firm, or to establish." We find this same word documented in various forms *(Amen, Amun, Amon,* or *Ammon)* among various modern-day African tribes such as the Mande, the Bantu, and the Yoruba and often specifically equated with the word *Amma.* We also find it among the sub-Saharan tribes that were roughly contemporary with ancient Egypt. Each culture defines the word with at least one of the two Dogon meanings, and in many cases, there is specific symbology associated with the related god to corroborate a link to the Egyptian god Amen (or Imn). There are also positive links to the same pronunciation of *Amen* or *Amun* with the same root meanings in Hebrew, Aramaic, and Greek. In other words, many cultures with likely claims of relationship to ancient Egypt are documented as pronouncing the word "Amen." However, in the prevailing view of the Egyptian hieroglyphic language, the name of the Egyptian god is pronounced "Imn" or "Emn." Budge defines both Dogon meanings with word entries pronounced "Amen,"[2] whereas the academically preferred *Altagyphtisches Worterbuch* by Adolf Erman, a German-based Egyptian hieroglyphic dictionary published in 1907, defines one as *Imn* or *Emn* and the other has been interpreted by some sources as *smn.* Also, we have seen that comparisons to Budge's word entries successfully link us with the Egyptian god Amen, whose symbology aligns with the African references. In essence, Budge has the African meanings and pronunciations right and the associated deity and symbology right, but related entries in *Worterbuch* fail to positively confirm either the Dogon god name or the second Dogon meaning. Yet common sense tells us that, based on cross-confirmed references from so many other related sources, the pronunciation of Amen should be seen as a correct alternate pronunciation for the name Imn.

It is not our intention here to rehabilitate Budge's dictionary. Rather, the inducement to choose Budge's dictionary as the basis of comparison for this study lies with Calame-Griaule's little-known French dictionary of the Dogon language, *Dictionnaire Dogon.* The Dogon speak a language called *Dogo so,* a name that, according to Dieterlen, means "Dogon word language."[3] This language was unclassified at the time of publication of *The Pale Fox* in 1965, partly because it includes upward of seventeen sublanguages. For example, the Dogon also possess a priestly language called *Sigi so,* the language of the *Sigui* ceremony, which is a secret language and includes far fewer words than Dogo so. According to Dieterlen, although words from the Mande and Gjur languages often make their way into Dogon speech, the Dogon in fact place a very high value on purity of language, because they consider language to be "on a par with knowledge."

Ironically, one of the key factors that compels us to use Budge's dictionary, rather than one of the more recent and preferred Egyptian language dictionaries, lies with Budge's own pronunciations of Egyptian words. Our starting point for these investigations is the language of the society that some may argue most closely resembles ancient Egypt—

that of the Dogon, a living culture whose priests are quite able to tell us how they pronounce their own words. Whenever two seemingly related Dogon and Egyptian symbols derive from common root concepts, such as might be represented by the Dogon word *sene* and the Egyptian word *sen,* we find Budge's pronunciations of the Egyptian words to be in consistent agreement with those defined in *Dictionnaire Dogon.* Likewise, when the Dogon language provides multiple meanings for a single word such as *Amen,* we often find that Budge defines the same set of multiple meanings for the same word, pronounced in a similar way. These correspondences are not theoretical or reconstructive in nature; rather, they are rooted in Dogon words *as they come out of the mouths* of living Dogon priests. Such correspondences are of very great importance to this study to the extent that it is a comparative study of two languages. And so Budge's dictionary becomes a tool of utility, providing a path by which to easily connect a given Dogon word to its likely Egyptian hieroglyphic counterpart. It is also worth noting that, in cases involving closely matched Dogon and Egyptian words that carry multiple common meanings, the suggestion is that the Dogon pronunciation provides one legitimate method of vocalizing the word and so may signal a need to reconsider some of the objections to Budge's pronunciations.

It is also worth noting that although Egyptian-language scholars may be inclined to evaluate the examples and word correspondences given here based strictly on linguistic criteria, there are other important factors at play that deserve consideration. The foremost of these is Dogon cosmology, which provides us with carefully documented definitions of many mythological words and symbols and which often outlines specific relationships between these words and symbols. And so, although in some cases the examples cited may cross the defined boundaries of two or more recognized Egyptian word groups, the connections we are exploring here should not be seen as purely linguistic ones; it may not always be sufficient to argue that two cosmologically related words do not derive from the same Egyptian root word. In these cases, a strictly linguistic interpretation may well be trumped by other equally legitimate considerations.

In the prevailing academic view of the Egyptian hieroglyphic language, relationships between words are defined by similarity of glyph construction. However, the Dogon do not have a written language; consequently, the Dogon priests define relationships between their words based purely on similarities of pronunciation. Likewise, there are signs that many of the words we will be discussing, such as *Amen,* may have preceded the appearance of written language in Egypt, so it is a safe conclusion that any preexisting relationships among these early Egyptian words could not have been founded on written glyphs. Dieterlen tells us that mere similarity of pronunciation is enough to arouse the suspicion of a relationship between words among the Dogon priests. From this perspective, an Egyptian word pronounced "amm" may be seen to link sensibly to a word pronounced "amma" without regard to the specific glyphs used to write the two words. Although this may be difficult for a mainstream Egyptologist to wholeheartedly accept, it constitutes business as usual among the Dogon.

In addition to publicly known common-usage meanings, the *Dogon Dictionnaire* often provides a second set of mythological meanings for a given word, meanings that are esoteric in Dogon culture and therefore known primarily to the Dogon priests. In several instances, *The Science of the Dogon* demonstrates, based on various Dogon drawings, that these esoteric meanings may also be supported by the shapes of the glyphs used to write related Egyptian words. What this implies is that there could well be a second level of meaning in the Egyptian language that is the counterpart to Dogon mythological meaning. In *Sacred Symbols of the Dogon,* when we encounter an Egyptian word that is pronounced like a Dogon word and that carries the same common-usage meanings as the Dogon word, part of our goal will be to examine whether the glyphs used to write the Egyptian word could also support the Dogon mythological meaning.

In most cases, the word entries in Budge's dictionary present several different spellings for the same word, and we often will be in the position of selecting one of these spellings for interpretation. We should bear in mind that Budge's dictionary presents us with what may be three thousand years of variant spellings of Egyptian words (consider how differently Shakespeare spelled words in English a mere four hundred years ago), and it is not our goal here to be all things to all spellings. Rather, we are trying to ask a more simplified question: Can the spelling of this word support a given symbolic interpretation? This question is best answered in terms of glyphs whose symbolic meanings we already understand. So as we begin to build a lexicon of Egyptian glyphs and related symbolic meanings, our ability to understand variant spellings of Egyptian words is likely to grow.

It is important to be aware that the opinions expressed by Budge regarding the evolution and structure of the Egyptian language do not always reflect the current thinking of modern Egyptologists. For instance, recent evidence does not support Budge's presumption of a purely ideographic or pictographic phase of the Egyptian language; rather, it seems to imply that the phonetic values of glyphs played a role from the earliest appearance of the language. It should also be noted that some of the word entries in Budge's dictionary actually represent prefixes and suffixes to words, not true words that could stand on their own within a sentence. For example, in English, the word *preverbal* refers to the time in life before a child learns to speak. The prefix *pre-* implies the concept of "before" but could not be used on its own as an actual word. Although such entries are not truly words, Budge's definitions of their meanings can shed light on our interpretations of Egyptian glyphs and of words shared commonly by the Dogon and Egyptian languages. For that reason, such word fragments will be examined and interpreted in this study on the same basis as actual Egyptian words.

There may well be a call for a more specific justification of method when it comes to the use of Dogon myths and language as a basis of comparison with their Egyptian counterparts. This will be particularly true in regard to the traditional Egyptologist, whose main reference to the Dogon may come from popular literature rather than a conversant familiarity with the anthropological studies of Griaule and Dieterlen—studies that are still held in the highest regard in the authors' native France. Let me first say that the similarities exhibited by the Dogon culture to that of ancient Egypt are far, far more than superficial; in fact, it would be more precise to say that the resemblances are pervasive. In support of that view, I will enumerate what I consider to be the main areas of resemblance:

1. The Dogon share many cultural and civic traditions with ancient Egypt.
2. Dogon religious traditions often reflect those of ancient Egypt and Judaism.
3. Dogon astronomical practices parallel those of ancient Egypt.
4. Dogon mythology is organized like Egyptian mythology. It includes similar stories about similar characters who often carry similar names and attributes and who play parallel roles in each mythology.
5. The Dogon esoteric priestly tradition is reflective of what we know about the priesthood in ancient Egypt.
6. Resemblances between Dogon and Egyptian words are not incidental. Most of the Dogon and Egyptian word comparisons presented here are based on similar pronunciation and at least two levels of common meaning, and they may be supported by common diagrams (Egyptian glyphs and Dogon cosmological drawings) or by parallel placement within the overall cosmology. This is not to say that every word in the Dogon language relates to an Egyptian word, but rather that the subset of Dogon priestly cosmological keywords seem to relate predictively to Egyptian counterparts.

7. These same Dogon words and drawings bear strong resemblance to those of the Amazigh, the hunter tribes who preceded the First Dynasty in Egypt and are considered the predecessors of the Berbers.

To evaluate the Dogon system in relation to the Egyptian, the traditional Egyptologist may be asked to tentatively set aside some of the hard and fast preconceptions of his or her profession and look at the Egyptian system through new eyes. If we think of the Dogon and the Amazigh as cultural stepsisters of ancient Egypt, we may not always find the glass slipper of modern Egyptology to be a perfect fit on the foot of the Dogon. However, when we reverse our view, we often cannot help but notice the near-perfect fit of the Dogon fur slipper on the foot of the Egyptian.

Although the impulse of the traditional Egyptologist may be to attribute Dogon/Egyptian resemblances to modern rather than ancient contacts, this view is not upheld by established facts about the Dogon, which again and again induce us to align Dogon culture with early Egypt. For example, the Dogon tradition preserves cosmological drawings that often resemble the shapes of Egyptian glyphs but that have not been organized in the form of a written language. Dogon cosmology includes eight ancestors with the same attributes as the eight paired Egyptian Ennead ancestor gods and goddesses, but without the assigned stature or names of actual gods or goddesses. The Dogon priests specify a method for building a pyramid-like aligned granary structure according to the same ancient plan as a Buddhist stupa, but not in the mature Egyptian form of an actual pyramid. The Dogon also practice many other traditions, such as circumcision, that are known to date from commensurably ancient times, and Dogon culture itself takes a form similar to that which was sustained in ancient Egypt for three thousand years—arguably the most stable social structure known to history. Therefore, a reasonable conclusion is that the Dogon, much like their Jewish cousins, managed to preserve these resemblances from a specific and self-consistent point in ancient times—the period just prior to the rise of the First Dynasty in Egypt and just prior to the appearance of written language. The geographic distance separating Egypt from Mali is not a compelling argument against this notion because the Dogon could easily have moved (in the course of five thousand years) a distance comparable to that moved by the American settlers in less than one hundred years. Nor can the resemblance of occasional Dogon words to pedagogically "later" Egyptian words be used as an argument against this view, because virtually any Egyptian word could theoretically date from a period earlier than its earliest known appearance in a currently known surviving text. From this perspective, we would expect to find many similarities between Dogon symbols and language and those of the Amazigh, and it is reaffirmingly true that we find just such resemblances.

As one final note, throughout this volume there will be many occasions in which we wish to illustrate Dogon and Egyptian symbols and related scientific diagrams. For the sake of convenience and consistency and as a way of underscoring the many intimate similarities that persist among these three symbolic systems, whenever possible these illustrations will be presented in the form of Egyptian glyphs.

TWO

SCIENCE AND THE STRUCTURE OF MATTER

Before we begin to discuss the relationship of mythological symbols to the structure of matter, we should take a few moments to review the structure of matter as science knows it. Today, we learn about the component elements and forces of matter through the science of astrophysics, which is the study of the universe and how it was formed. Astrophysicists formulate theories, conduct experiments, and use mathematical models to infer what the universe might have looked like at the time of its formation and what it might ultimately look like at the time of its demise. Astrophysics includes the study of the major concepts and forces that govern the universe, such as light, time, mass, gravity, and acceleration, as well as how these concepts and forces interact. The structure of matter itself is also the domain of the astrophysicists, from the point of its primordial conception to its ultimate emergence in the familiar form of matter as we know it.

The prevailing scientific theory used to describe the formation of the universe is the big bang theory, which postulates that the universe started out as an infinitely condensed ball of matter. Some undefined impulse caused this ball to rupture in an explosion of superhot primordial matter, which spun and cooled and eventually formed all of the spiraling galaxies of stars and planets. The force of this initial explosion caused the universe to expand, growing ever larger in size over time. One scientific school of thought originally held that the universe would expand infinitely. Another school proposed that the gravitational effect of matter would eventually slow the growth of the universe and then perhaps ultimately reverse it, causing matter to shrink back down to its original, infinitely compressed size. More recent measurements by astrophysicists seem to show that the rate at which the universe expands is actually increasing, which suggests that the expansion rate might be accelerated by some unknown or undefined force. (Other researchers such as the late Eugene Mallove have begun to sharply question the very scenario of the big bang, citing new research into fields such as cold fusion.)

What we know about the structure of matter begins with atomic theory, whose purpose was to define the atom—the fundamental building block of matter—and its subcomponents. Atomic theory describes the internal structure of the atom itself, which consists of protons, neutrons, and electrons. Its rules also apply to each of the diverse elements that compose the Periodic Table of Elements, such as hydrogen, oxygen, nitrogen, and many others. Atomic theory tells us that an electron orbits around the nucleus of an atom, which is made of protons and neutrons, and it diagrams the various shapes that an orbiting electron can trace. It also describes how an electron can be shared between two atoms or can be stolen from one atom by another to form a molecule such as water.

The forces that bind the particles of an atom together are called quantum forces and are discussed by another scientific theory called quantum theory. Quantum theory includes four quantum forces: gravity, which causes bodies of mass to be drawn together; electromagnetic force, which causes an electron to orbit the nucleus of an atom; weak nuclear force, which binds smaller particles called quarks together to form protons and neutrons; and strong nuclear force, which binds the components of the nucleus of an atom tightly together. There are many different kinds of quarks. So far, more than two hundred different types of quarks have been identified. Quantum theory also describes the rather unusual behavior of these quantum particles and forces. In fact, by the time we reach the level of matter described by quantum theory, matter behaves in ways that might seem quite bizarre when compared to our everyday experience.

These two theories of matter—atomic theory and quantum theory—have been tested experimentally and so are well-grounded in fact. However, the science of astrophysics still includes major elements, such as the big bang theory and string theory, whose details are not yet fully proved and so are still very much open to debate. String theory is an approach to the study of astrophysics that emerged in the 1980s—a full quarter of a century after Griaule and Dieterlen's studies among the Dogon. String theory proposes that all matter is derived from the vibrations of tiny one-dimensional strings. The various vibratory patterns of these strings are compared to the range of notes played on a violin—each of which produces its own distinct, recognizable tone. In turn, the various vibrational patterns are thought to produce what we see as the many diverse particles of matter.

Early in the history of string theory, several different versions of the theory emerged, each with its own unique outlook on how matter might be formed. Eventually, string theorists realized that these seemingly independent theories actually described different aspects of one broader view of the formation of matter. Many of the differences between these versions were later reconciled by a single unifying theory called M-theory, which serves as a kind of umbrella over all of the other versions. For the purposes of this study and for reasons that will become clear through discussion, we will assume for the sake of argument that the details of the big bang theory and M-theory are substantially correct and therefore allowable as a fair basis of comparison with the descriptions of the Dogon.

If the big bang theory is correct, then as we have already said, our universe began billions of years ago as a single, unbelievably dense ball of mass, containing all of the potential matter of the future universe. In his book *A Brief History of Time,* Stephen Hawking compares the unformed universe to a black hole—an astronomical body so very dense that even light cannot escape from its gravitational pull. Because no light leaves a black hole, it has no reflected image in the traditional sense, so Hawking defines its shape as the boundary formed by the path of light rays that are unable to escape from the hole—what scientists refer to as the event horizon of a black hole.[1]

The event horizon of a black hole

When this ball of mass—the precursor of the future universe—opened, it released a superheated brew of potential matter that astrophysicists call quark/gluon plasma. This plasma cooled over time, and as it cooled, it passed through what are known in science as phase transitions, such as the processes by which water vapor condenses into water or liquid water freezes into ice. These transitions ultimately resulted in the formation of familiar matter, starting with the simplest and most abundant element in the universe, hydrogen. Because each hydrogen atom includes only one electron, one proton, and one neutron, and because matter seems inclined to migrate toward forms that include pairs of electrons, hydrogen atoms tend to bond together in twin pairs to form a molecule identified in chemistry as H_2.

Astrophysicists define light as one of the key limiting factors of the universe. Light travels at a rate of 186,000 miles per second, and Einstein's famous special theory of relativity demonstrated that all observers—no matter what their own relative speeds—must measure the same speed of light. This idea runs contrary to common sense, which is based on our experiences interacting with everyday objects. For instance, if you were traveling in a car moving at 10 miles per hour alongside a train, which was itself traveling at 50 miles per hour, you would measure the train's speed relative to yours as 40 miles per hour. However, this mathematical pattern does not hold true when we speak about measurements of the speed of light. According to Einstein, no matter how fast a person moves, the speed of light must remain a constant 186,000 miles per second relative to the person observing it. Because speed is calculated simply as distance divided by time, for this to be mathematically possible, time must slow down as acceleration increases. Einstein's theory, which is based on the famous formula $E = mc^2$, defined a dynamic relationship between acceleration, time, and the speed of light that implies that an accelerated body can never reach the speed of light.

Another limiting factor in the universe is the force of gravity. Einstein's special theory of relativity was meant to apply only to bodies acting outside of the influence of gravity and did not attempt to address it as a force. One purpose of his general theory of relativity was to incorporate the effects of gravity, a constraint that created difficult theoretical problems for Einstein. His solution to these problems was to define gravity not as a force, but rather as an effect caused by the shape of a kind of invisible fabric of the universe that he called space-time. Einstein suggested that we think of space-time as a stretched bed sheet and that we imagine a body of mass—such as a planet or a star—as a basketball placed in the center of that sheet. No matter how tightly we might stretch the bed sheet, the basketball will create a dimple in the fabric—similar to the kinds of dimples Einstein proposed must be created in the fabric of space-time by bodies of mass. The shape of these dimples would deflect the motion of other, smaller bodies as they move through space and cause them to circle around the larger bodies. The more massive the body, the greater the bending or warping of space-time would be around it, creating a greater gravitational pull related to that body. In the world of physics, the effects of acceleration on a body are equivalent to those caused by an increase in its mass, and so if time must slow down for an accelerated body, it must also slow down for a body with increased mass.

In the view of quantum theory, all forces are thought of as if they were caused by particles. For instance, gravity is said to be caused by a very subtle particle called a graviton—so subtle, in fact, that one has never been detected. Quantum theory defines more than two hundred types of quarks and groups these particles together into four categories based on a property called spin. The concept of spin might make you think of a top spinning on its axis, but in truth, these tiny particles have no well-defined axis to spin around. Rather, what the quality of spin actually tells us is what the particles look like from different directions. A particle that looks the same when viewed from any direction is said to have a spin value of 0; one that must be turned in a complete circle to look the same has a spin value of 1; a particle that must be turned half-way around to look the same has a spin of 2; and a final, very odd set of particles, which actually must be turned around *twice* to look the same, have a spin of ½. Again, our experiences in a three-dimensional world would tell us that a particle should never have to be turned around twice to look the same, so clearly there must be something unusual going on at the subatomic levels of matter that causes these particles to require an extra turn.

Anytime we examine components of matter as small as quarks, we must be aware that their tiny size can have an effect on the outcome of our examinations. The tools we have at our disposal are so very large when compared with these miniscule particles that our efforts to examine them are somewhat like trying to examine a flea with a sledgehammer. There is a tenet of quantum theory called Heisenberg's Uncertainty Principle that states that, when examining very lightweight particles, the very act of perceiving them will necessarily disturb them.

In practice, what this means is that we cannot simultaneously know with complete accuracy both the position and the momentum of one of these particles.

From the beginning, one of the great confusions of quantum theory involved the behavior of these tiny particles. Certain experiments indicated that, on a quantum level, matter behaved like waves of water. Other experiments clearly showed that such matter behaved like particles. There was great confusion at first as to which was the correct point of view: Did matter at the quantum level behave like particles or did it behave like waves? Careful experiments eventually showed that both answers were correct. Matter, prior to being observed or detected, behaves like waves; after being detected, it behaves like particles. One of the great mysteries of quantum science, yet to be solved, is how the mere act of perception can cause such a fundamental change in the behavior of matter.

Another great confusion about matter in its primordial, wavelike state involves the inability of astrophysicists to predict where a particle will ultimately appear when it is perceived. To all outward appearances, an unperceived particle in its wavelike state seems to exist everywhere at once. Some researchers describe these particles as being "smeared out" in their wavelike form. The ultimate point of emergence of a perceived particle seems to be entirely unpredictable. This apparent randomness in the process of the formation of matter was a source of great upset to Einstein, who felt on an instinctive level that the formative processes of nature could not be based on ultimate randomness.

Yet another unexplained attribute of matter is its dual nature. The particles and forces of quantum theory come in paired sets; likewise, larger particles of matter—protons, neutrons, and electrons—also tend to form pairs. No current theory of astrophysics provides a complete explanation for this dual nature of matter. As matter is created, what force, attribute, or function causes it to form pairs, and at what stage in the process does this dual nature manifest itself?

In the 1980s, string theory offered a new outlook on the possible nature of matter. It proposed that all of the various subparticles and forces of matter are really just by-products of the vibrations of tiny, one-dimensional loops of matter called strings. Along with the idea of these vibrating strings came new insights into the dimensional nature of matter. In the earliest days of astrophysics, the universe was thought to have three dimensions: height, length, and depth. Einstein's theory of relativity added a fourth dimension—time—to the original three. Then, string theory proposed to add an additional seven unseen dimensions to these four. According to string theory, these dimensions never fully opened at the time of the formation of the universe but instead exist in wrapped-up bundles at every point of space-time. These bundles are called Calabi-Yau spaces, after the last names of their two discoverers, Eugenio Calabi and Shing-Tung Yau. Brian Greene suggests in his book *The Elegant Universe* that we can conceptualize how these wrapped-up dimensions work by imagining an ant walking across a distant power line. From a distance, it might seem that the ant can only move in two directions—forward or backward along the length of the thin power line. However, if we examine the ant more closely, we see that it actually can move in a third, previously unseen direction—in a circle around the circumference of the power line.

According to string theory, strings vibrate in these seven unseen dimensions and then tear space-time to form the next Calabi-Yau space. The formulas of string theory allow for more than two hundred thousand possible shapes for the Calabi-Yau space, each with its own set of potential attributes for a possible universe. String theorists are presently using mathematical models to analyze these many possible shapes in an effort to determine which ones might be capable of producing a universe with properties such as the ones we know. String theorists have defined three types of string intersections, which were described in the previous chapter. M-theory also predicts that strings have a tendency to combine together to form membranes. A membrane might exist in two dimensions, much like the thin membranes of the human body, or it might exist in multiple dimensions. So string theorists refer to each type of membrane specifically in terms of its number of dimensions. A two-dimensional membrane is called a 2-brane, a membrane that exists in three dimensions is called a 3-brane, and so on.

If we were to consider components of matter as they are defined by these scientific theories, organize them roughly in sequence by their order of appearance, and present them in the form of a simple list or table, it might look like this:

SCIENTIFIC STRUCTURE OF MATTER
Unformed Waves
Perception
Vibration/Mass
Calabi-Yau Space
String Intersections
Membranes
Quarks
Protons/Neutrons/Electrons
Atoms

As our understanding of the formative components of matter evolves, we can see that there are few aspects of its structure that would be intuitively obvious to a person based on everyday experience. There is nothing in our daily world to suggest that primordial matter should behave like waves and particles, nothing that would lead us to think that matter is formed from tiny vibrating threads. In fact, Einstein himself had no cause to believe that matter was formed from threads. Likewise, there would be no obvious reason to surmise that these threads pass through seven vibrational patterns as part of their journey to create matter. Nonetheless, *The Science of the Dogon* shows that each of these attributes is explicitly defined and diagrammed as part of the Dogon mythological structure of matter. In fact, the Dogon actually describe *each* of the constituent components of matter—organized in the correct sequence—as part of their complex cosmology. So the natural impulse would be to wonder whether the Dogon myths might somehow have had an original basis in science.

THREE

DOGON COSMOLOGY

The stated purpose of Dogon cosmology—clearly defined by the Dogon priests themselves—is to describe the creation of the universe and matter by the god Amma. The events of this mythological creation are revealed through a carefully composed set of story lines whose role is to introduce the concepts, themes, keywords, and symbols that are the foundation of Dogon cosmology. These myths work on two levels at the same time—one public and one private. The public myths are more general in nature; their function is to describe key mythological episodes, introduce major themes such as the concept of water as a source of creation, and establish recurring mythological symbols such as the serpents, spiraling coils, and clay pots that are so familiar to us from various world mythologies. The private myths operate on a more discreet level; their details are reserved for initiates of the Dogon religion. The role of these myths is to define the various elements of Dogon cosmology, first by providing explicit descriptions for each individual component of the cosmology. Many of these components are associated with mythological keywords—carefully couched words and phrases expressed in the terms of Dogon language. These words also may carry two levels of meaning—one common-usage meaning pertaining to everyday speech and one mythological meaning pertaining to the word's larger symbolism within the cosmological structure. When possible, a component may be further defined by its relationship to a cosmological drawing or ritual object whose image and associated symbolism increase our knowledge about the role of the component within the larger cosmology.

Dogon cosmology begins with the god Amma, who was the creator of the universe, and with Amma's egg, a primordial body defined as having existed before the creation of the universe and said to have housed all of the potential seeds and signs of the future creation. We know from Dogon tribal drawings and other physical renderings that Amma's egg looks like an inverted cone. In fact, images of Amma's egg strongly resemble a scientific diagram of the event horizon of a black hole—the astronomical body that scientists say most closely resembles what the unformed universe may have looked like.[1] The Dogon myths tell us that within this primordial egg, Amma's motion forms a spiral, which the Dogon refer to as an accelerated ball. When Amma broke the egg of the universe, it released a whirlwind that spun and scattered primordial matter in all directions—matter that would eventually come to form all of the galaxies, stars, and planets. The myths say that the planets were thrown out like pellets of clay and describe the sun as a clay pot raised to a high heat. In one mythological episode, we are told that the sun is surrounded by a spiral with eight turns—a statement that anticipates the eight separate solar zones identified by modern science. The myths correctly describe the moon as a dead, dry body and reaffirm the assigned symbolism of clay in relation to heavenly bodies by comparing it to dried clay.

The shape of Amma's egg

Dogon cosmology tells us that the universe is based upon a principle of twin births but that the process of creation began with a flawed union between Amma and Earth, an incestuous act that resulted in the birth of only one being, the Jackal. This birth was seen as a breach of order in the universe, and so the Jackal became a symbol, both of disorder and of the difficulties of Amma. The Dogon say that water is the divine seed that entered the womb of the earth, fertilized it, and produced the perfect twin pair—the *Nummo*—a word that is synonymous with water in the Dogon language. Just as eight spirals are said to surround the sun and give it its fundamental movement, the Dogon say that the spiral of the mythological Word of Amma gave the womb its generative powers.

The first finished work of Amma to emerge from the egg of the universe was a tiny seed called the *po*—a structure that, like the atom, is considered to be the primary building block of matter. According to Dogon mythology, the po is the very image of the creator; Amma's creative will is said to be located inside the po. The myths also say that the po is the image of the origin of matter—that all matter is formed by the continuous addition of like elements, beginning with the po. These outward similarities between the po and the atom are reaffirmed by mythological descriptions of the po's inner structure. Dogon cosmology describes the po as being comprised of an even smaller class of elements called *sene* seeds, which combine together at the center of the po and then surround it by crossing in all directions to form a nest. Dogon descriptions of the behavior of the sene follow closely with those of atomic theory and thereby enable us to identify the sene as mythological counterparts to protons, neutrons, and electrons. Dogon mythology even provides us with a drawing of the sene that takes the form of a four-petaled flower—the same essential shape as one of the typical orbital patterns traced by an electron as it encircles the nucleus of an atom.

Shape of the sene seed

Scientific descriptions of protons and neutrons tell us that they are each composed of three quarks. We also know that quantum experiments have revealed more than two hundred different types of quarks. These statements are in close agreement with Dogon descriptions at the same component level of matter, although the Dogon descriptions can often be substantially *more specific* than their scientific counterparts. The Dogon myths state that sene seeds are "germinated" from 266 fundamental seeds or signs—seeds that can be seen as the mythological counterparts to quarks in quantum theory. We know that scientists categorize quarks based on a property called spin, a term that essentially tells us the rotational properties of the particle. The Dogon mythological drawing that represents the germination of the sene consists of four spiked, circular figures, each of which reflects (to the extent possible in a two-dimensional drawing) the rotational attributes of one of the four spin categories.[2] We may recall that quantum theory defines four quantum forces—gravity, electromagnetic force, weak nuclear force, and strong nuclear force. These forces correspond to the four branches of a mythical Dogon seed called the *sene na,* in which the formation of matter is said to take place. The names of these four branches in the Dogon language carry meanings that correspond to each of the four quantum forces (see the table below).[3] The mythological names of the three branches that correlate to forces of astrophysics that directly affect the binding of components of the atom—protons, neutrons, and electrons—all include the word *sene* in their names.

BRANCHES OF THE SENE NA

Dogon Branch	Meaning of Dogon Name	Corresponding Quantum Force
Mono	"To bring together"	Gravity
Sene gommuzu	"Bumpy"	Electromagnetic force
Sene urio	"That bows its head"	Weak nuclear force
Senu benu	"Stocky"	Strong nuclear force

As we work our way downward through the component levels of matter, we move out of the realm of quantum theory and into the theoretical domain described by string theory or torsion theory. We recall that string theory defines matter as the end product of tiny, vibrating strings. Although many of the specific details of string theory have not yet been exhaustively proved, we find that the parallels presented by Dogon cosmology do not fail us in regard to string theory. Rather, the Dogon myths specifically confirm what string theorists propose. The myths tell us that the formation of matter begins with tiny, vibrating threads woven by a mythological spider named Dada. As in string theory, which describes more than 200 quantum particles, these threads are said to be the source of the 266 primordial seeds or signs that combine to form larger components of matter that seem to be counterparts of protons, neutrons, and electrons. In fact, one of the mythological drawings of the Dogon, which is presented in the form of an annual field drawing, is specifically meant to represent these seeds or signs. This drawing takes a form similar to one of the documented vibratory patterns of a string in string theory.[4]

If we follow the components of string theory upward through the structure of matter, we arrive next at the Calabi-Yau space—the seven wrapped-up dimensions within which the string is thought to vibrate. At the corresponding level of matter, Dogon cosmology describes the formation of an egg that is the mythological home to seven successive vibrations of matter. The Dogon define and diagram these vibrations as seven rays of a star, each of increasing length. Although string theorists often describe strings as loops, the Dogon conceptualize their mythological thread as a coil, the shape of which is based on the spiraling figure that can be traced along the endpoints of the seven starlike rays inside the mythological egg.

Shape of the Dogon thread

Descriptions from string theory tell us that after the seventh vibration of a string, a Calabi-Yau space tears and then bends in a new way to form another Calabi-Yau space. Likewise, Dogon cosmology states that the vibrations inside a mythical egg can progress only after the thread has "broken through" the wall of the egg. This event is defined both as the eighth and final stage of the first egg and as the initiating stage of a new egg.

Vibrations within the Dogon egg (Calabi-Yau space)

One of the stated purposes of the Dogon myths is to introduce and define important mythological concepts and symbols. These symbols take many different and sometimes subtle forms. They begin with images of spiraling coils as a formative shape within the processes of creation. Story lines within the myths encourage us to associate images of clay and clay pots with various astronomical bodies formed within the universe, including the sun, the stars, the planets, and the moon. They establish the numbers 2, 7, and 8 as key cosmological numbers that repeat again and again throughout the story lines of the myth. They place emphasis on key mythological shapes, many of which are presented later in the form of cosmological drawings. They define the art of weaving as an important mythological concept and relate it to the formation of matter, and they associate the growth of matter with component elements and stages of agriculture, such as seeds, flowers, and the growth of plants.

The Dogon myths dedicate much discussion to the concept of the granary, a structure that is shaped like a flat-topped pyramid with a round rather than a square base.[5] Within the Dogon mythological system, the granary represents a unit of volume, and much attention is paid to the configuration and dimensions of this structure. The Dogon say that the granary provides examples of many of the key geometric shapes, while at the same time it conveys important astronomical symbolism. The round base represents the sun, and a circle within the square top symbolizes the moon. The implication is that the granary itself, which stands between the symbolic sun and moon, represents the concept of Earth. The structure of the granary is roughly pyramidal; its overall shape recalls that of a squared hemisphere like the Great Pyramid, whose height bears the relationship of pi to the perimeter of its base. The myths include very specific dimensions for the granary. It is defined as having eight internal chambers, each associated with an important agricultural grain. The granary is also defined to include four staircases—one up the middle of each flat side. Each staircase is associated with one of the four

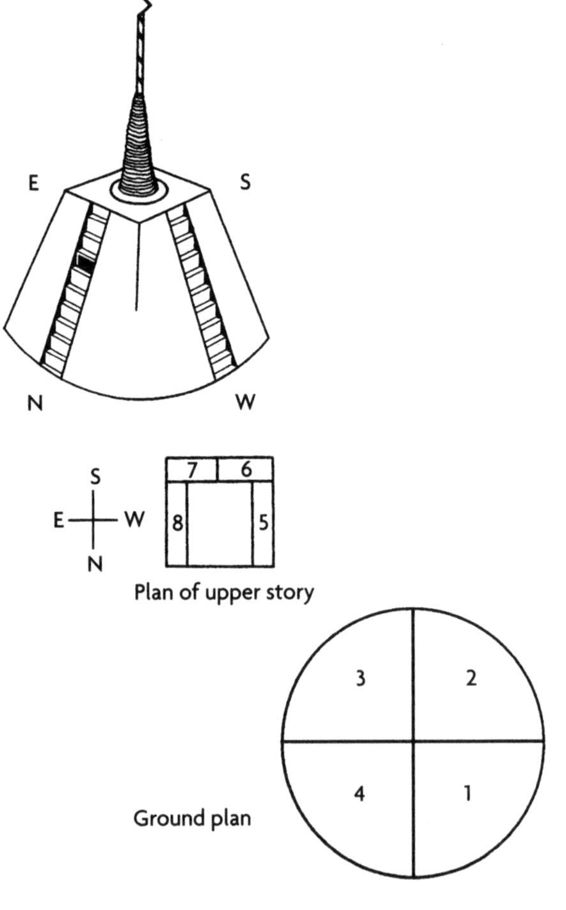

The Dogon Granary
(from Griaule, Conversations with Ogotemmeli, *33)*

key constellations that govern the agricultural cycle and with a zoological class of animals or plants. The myths define ten steps in each staircase; the rise and tread of each step is interpreted to represent a different order or phylum of the animal or plant kingdom.

Dogon mythology also defines a four-step creation process whose stages are meant to apply to any creative act of man or god. The first stage is called *bummo,* and it is considered to be the stage in which a project is conceived. The next is called *yala;* at this stage the project is defined in broad strokes or *signs.* The third stage is called *tonu,* and it is at this stage that the image of the project is revised or refined. The final stage is called *toymu* and represents the concept of completion.

If we were to organize the component stages of matter as described by the Dogon cosmology and present it in roughly chronological sequence, the following table would be the result:

DOGON MYTHOLOGICAL STRUCTURE OF MATTER
Waves of water
Vibrating threads
Seven vibrational patterns within the egg (Calabi-Yau space) 
The coiled thread
Four quantum forces
Germination of the sene (More than 200 quarks/266 seeds or signs)
The sene seeds (protons, neutrons, electrons) 
The po (atom)

By now, we may have lost count of the number of specific points of correspondence between the scientific and the Dogon mythological structure of matter, even as presented within the scope of this brief overview. However, we can see that the two descriptive structures of matter—scientific and mythological—run parallel to each other through level after component level of matter. The Dogon myths describe and diagram the stages of matter correctly according to science—organizing their components in a proper sequence, assigning appropriate attributes to each component force of matter, and providing us with many necessary and correct corroborating details, including specific drawings of several key components. The very least we can say about these myths is that they accurately describe what they profess to describe—the structure of matter and the creative forces and elements that comprise it.

FOUR

DOGON SYMBOLS AND EGYPTIAN GLYPHS

In *First Steps in Egyptian Hieroglyphics,* Sir E. A. Wallis Budge expresses his belief that the Egyptian glyphs were originally conceived as pictographs or ideographs—drawings whose purpose was to convey a specific image, idea, or concept. He believed that phonetic values came to be associated with some glyphs at a later point, but sometime very early in the history of the language. More recent studies suggest that phonetic values may have been an original attribute of the Egyptian glyphs and that these phonetic values made their appearance during the formative period of the Egyptian state. Budge believes that, on a fundamental level, the meaning of an Egyptian word was originally determined simply by combining the concepts depicted by the glyphs used to write it, much as you might solve a child's rebus puzzle, only in this case substituting conceptual rather than phonetic values for the glyphs. Budge cites examples of early Egyptian words whose meanings seem to have been established in precisely this way, such as the word *ai*,[1] which means "to cry out," a meaning that is reflected in both the glyph structure and pronunciation of the word. He goes on to tell us that other ancient languages that date from the same time period also exhibit a pictographic or ideographic aspect. Such languages include the Sumerian language, whose ideographs eventually evolved into cuneiform characters, and the Chinese language, which some researchers feel retains evidence of its ideographic origins even to this day.

Egyptian glyphs could also be used in the role of determinatives—characters whose purpose was to establish the context of the word. Perhaps the most obvious use of a determinative is in the appending of a god or goddess glyph to the end of an Egyptian word to identify it as the name of a deity. Other glyphs that appear at the ends of Egyptian words are also considered by Egyptologists to be determinatives, but in some cases, the rationale behind the inclusion of the glyph is more difficult to explain. So from one perspective, the glyphs represented simple phonetic characters whose purpose was to convey a specific sound, much like the letters of the English language. From another, they represented concrete representations of known objects or concepts. And from a third, they were used as determinatives to establish context.

Just as the esoteric Dogon and Egyptian traditions are known to have had their public and private aspects when it comes to mythological story lines and keyword meanings, it also makes sense that they might present public and private definitions for key glyph shapes. Also, it is within a public context that we understand the assignment of both the hemisphere glyph to mean a loaf of bread and the wave glyph to mean water as well as the likely derivation of the phonetic values of these glyphs from the names of the objects they depict.

My introduction to the Egyptian hieroglyphs came through an indirect path, by way of Dogon myth and language. Long before I seriously perused my first Egyptian glyph, I was well acquainted with the symbols of Dogon mythology. I knew based on Ogotemmeli's drawings and descriptions that in Dogon culture, the shape stood for the unformed universe. Likewise, I was aware of the meanings of several other key shapes from Dogon mythology, each of which closely resembled an Egyptian glyph. When my comparisons of Dogon and Egyptian culture finally led me to the Egyptian language, these shapes and meanings were already familiar to me, and so I naturally was inclined to apply the Dogon meanings to similarly shaped Egyptian glyphs. It was my impulse, based on knowledge of Dogon culture and language, to interpret the Egyptian glyphs as pictographic or ideographic placeholders for larger concepts—much as Budge says they originally may have been read. Likewise, my inclination was to interpret the Egyptian words by substituting larger concepts for each pictographic or ideographic character and then simply reading out a meaning, much as you might substitute the full phrase "National Biscuit Company" for the acronym NABISCO in English. My experience with Dogon myths and symbols also led me to associate the pronunciations of Dogon mythological keywords with corresponding glyph shapes. This method of interpreting Egyptian words is illustrated by various examples in *The Science of the Dogon,* which uses keywords and descriptions from Dogon mythology to reveal the likely meanings of Egyptian glyphs.

The Dogon myths work to establish meaning through stories, key words, symbols, structures, rituals, and drawings, and they also make careful use of recurring themes as a way of defining important concepts. Discussions in *The Science of the Dogon* illustrate how each of these conventions is used to reaffirm the likely scientific meanings of a handful of mythological shapes, each of which matches a specific Egyptian glyph and carries a meaning that is supported by its own explicit references. The relationship between these shapes and the corresponding Egyptian glyphs provides us with a kind of "starter set" of glyph characters whose symbolism we believe we understand because their definitions are drawn directly from Dogon mythology.

On the next page is the first of a series of tables of Egyptian glyph shapes, intended to document for the reader the likely meanings of those shapes as I interpret them and to identify the source from which those meanings have been drawn.

DOGON MYTHOLOGICAL SYMBOLS AND MEANINGS

Glyph	Shape Description	Symbolic and Scientific Meaning
	Unformed universe	In Dogon cosmology, this glyph represents Amma's egg, the primordial structure said to have held the potential seeds and signs of the future universe. It takes the same shape as Stephen Hawking's diagram of the event horizon of a black hole (see Hawking, p. 86; and Scranton, p. 57).
	Crossing of the sene seed	A Dogon drawing with this shape is used to represent the path of the sene seed, one of the formative components of the po, which in Dogon cosmology is the mythological counterpart to the atom. It matches one of the classic shapes of an electron orbit in science (see Scranton, p. 67).
	Spiraling coil	This glyph takes the same shape as the cosmic thread defined in Dogon cosmology. It is the primordial body whose vibrations result in the formation of matter. It corresponds to a string in string theory (see Scranton, p. 77).
	Clay pot	Dogon mythology associates images of clay and clay pots with bodies of mass such as stars, planets, and moons. From the standpoint of Dogon cosmology, a clay pot seems to represent a particle (in the sense of particles and waves).
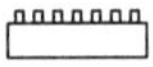	Drawing board or weaving loom	In everyday usage, this glyph represents a drawing board or weaving loom. From the standpoint of mythology, it seems to represent the strong nuclear force, which binds particles together tightly in the atom. In Dogon mythology, it is said to "draw the atom." It would correspond to the fourth branch of the sene na seed, whose name means "stocky" (see Scranton, p. 78).

Glyph	Shape Description	Symbolic and Scientific Meaning
⌓	Hemisphere glyph	The shape of the hemisphere glyph most closely resembles the shape of the Dogon granary, which Ogotemmeli said represents earth. The mythological keyword "earth" seems to correspond to the scientific concept of mass or matter. The eight chambers of the Dogon granary recall the eight stages of the Calabi-Yau space (the Dogon define seven stages of vibration and an eighth stage of tearing, during which the coiled thread attains mass).

From *The Pale Fox*, *Conversations with Ogotemmeli*, and *The Science of the Dogon*.

Budge's *An Egyptian Hieroglyphic Dictionary* provides us with several methods by which to learn the meanings of individual Egyptian glyphs. The most obvious source of meaning is Budge's List of Hieroglyphic Characters, which is located at the front of his dictionary. There, he presents 1,490 Egyptian glyphs, their pronunciations (in the prevailing academic view, only a relative handful of Egyptian glyphs carry direct phonetic values), a description in English of the glyph (if known), and often a second definition or phonetic reference expressed entirely in terms of other Egyptian glyphs. In some cases, the definition of a glyph may be unknown or uncertain. Budge clearly indicates in these entries if he is uncertain about the meaning of a glyph. From this list we often learn which object, idea, or concept he feels is depicted by the glyph and the common-usage meaning of the glyph in the Egyptian hieroglyphic language.

The following table summarizes individual glyph meanings as Budge identifies them in his "List of Hieroglyphic Characters," which is found at the front of his *An Egyptian Hieroglyphic Dictionary.*

MEANINGS OF EGYPTIAN GLYPHS
From Budge's List of Heiroglyphic Characters

Glyph	Meaning	Budge Page Reference
	To bow	p. xcviii
	To speak	p. c
	To hide or conceal	p. ci
	To see	p. cvi
	Any act or force done with the arm	p. cvii
	Jackal-god, judge	p. cxi
	Tree or wood	p. cxxi
	Time	p. cxxi
	Shining light	p. cxxv
	The underworld (Tuat)	p. cxxv
	Land, earth	p. cxxi
	To stand up	p. cxxx
	The blade of an adze (woodworking tool)	p. cxxxix
	Sign of the dual	p. cxlvi

From *An Egyptian Hieroglyphic Dictionary.*

In addition to this list of glyphs, each section or chapter of Budge's dictionary—like any of the alphabetic sections in an English language dictionary—consists of words that begin with a specific glyph or glyph sound. There are often word entries at the beginning of these sections (and others within the body of the dictionary itself) that are written with a single glyph or a combination of just two glyphs. Also, in many cases there are Egyptian words that carry the same pronunciation as an Egyptian glyph, whose definitions may have bearing on the meaning of the glyph. These entries can provide additional insights into the possible meanings of a glyph.

MEANINGS OF EGYPTIAN GLYPHS
From Single-Glyph Word Entries

Glyph	Meaning	Budge Page Reference
	I, me, my, that which, that which is	p. 15a
	Place	p. 197a
	Building or structure	p. 438a
	To give, to set, to place	p. 864a

From *An Egyptian Hieroglyphic Dictionary.*

If, as scholars of the Egyptian language believe, the Egyptian glyphs originally carried ideographic meaning, then in cases in which Dogon or Egyptian mythology has already provided us with a likely definition for a full word, simple context can become a powerful tool for identifying the likely meaning of a component glyph within the word. This is especially true if the inferred meaning of the glyph matches its drawn image. In such cases, Budge agrees that the image of an Egyptian glyph can provide a high degree of self-confirmation of its own meaning.

In the English language, the letters of a word are written from left to right. The Egyptian hieroglyphic language provides for more possibilities than this. Glyphs can be written from left to right, from right to left, or even from top to bottom. When the text is arranged horizontally, we read the glyphs from the direction in which they face—for example, we read from the left and from the right. In Budge's *An Egyptian Hieroglyphic Dictionary,* and in the numerous examples presented in *Sacred Symbols of the Dogon,* the glyphs of a word are read as if in columns from top to bottom, starting at the left and moving to the right (note that to accommodate a particularly wide glyph, exceptions to this general rule will be seen).

1 3 5
2 4

When Egyptian hieroglyphs are read phonetically, some vowel sounds are not actually written but rather are inferred by the reader. This is also true of other early languages such as ancient Hebrew, which later in its history acquired markings to indicate vowel sounds. In its own way, the absence of written vowels in the Egyptian hieroglyphic language supports the belief that the language originally could be read pictographically or ideographically. In our earlier example of the word *ai* , which means "to cry out," the meaning of the word is made evident by the form of the glyphs used to write it, which can be read literally as "the act of crying out." Based on an ideographic reading, no vowel sound is required, so none is lacking. It is only when we impose a phonetic value upon the characters of a word that the vowel sounds appear to go missing.

Because the Dogon and Egyptian cosmologies share many root words in common, there is the possibility that we can use these common words to help establish the meanings of Egyptian glyphs. Likewise, whenever the Dogon myths provide a precise definition for a word, we may be able to look to similar Egyptian words for further clarification of meaning. In one example from *The Science of the Dogon,* comparisons of the Dogon word *sene* and the Egyptian word *sen* are used to confirm the identity of sene seeds as components of the atom. We recall that Ogotemmeli defined the sene as a class of components that comprise the po—the Dogon mythological symbol that closely resembles the atom. These seeds are said to first commingle in the center of the po and then cross in all directions to form a nest surrounding it, very much like the arrangement of protons, neutrons, and electrons in a real atom. Dogon mythology also provides a picture of the sene that takes the form of a typical electron orbit. So the meaning of the word *sene* is explicitly established in three different ways: from its position in relation to the mythological atom, from Ogotemmeli's careful descriptions, and from a specific drawing that takes the correct diagrammatic form. If we presume that the Egyptian door bolt or knotted rope glyph means "to tie or bind," then the mythological definition of protons and neutrons is confirmed by the Egyptian word *sen,* whose common-usage meaning Budge reports as "they," "them," or "their."

PROTON/NEUTRON (SEN/SENE)

They, them, their		The binding of a particle from three elements . (Protons and neutrons are each formed from different combinations of three quarks.) (See Budge, p. 603a; and Scranton, p. 94)

Discussions in *The Science of the Dogon* show that one cosmological meaning of the word *sene* also refers to an electron, the body whose electromagnetically induced orbit causes it to circle around the nucleus of the atom, forming what the Dogon describe (in common usage) as a nest. We can confirm the meaning of this Dogon keyword by looking to the root of the Egyptian word for nest *(aun)*. Our interpretation of this word depends upon the assignment of the meaning of "that which" to the reed leaf glyph , a value that is affirmed by word examples in

James P. Allen's *Middle Egyptian.*[2] These examples, , meaning "one who is in the heart," and , meaning "who/which is in," illustrate a meaning of "one who" or "[that] which is" for the reed leaf glyph. A similar meaning can be inferred based on the meaning of an Egyptian word that, for Budge, carries the same pronunciation as the glyph.[3] Based on this meaning, the definition of "an electron" becomes clear.

ELECTRON (AUN)

To open (nest)		That which orbits due to the electromagnetic force . (See Budge, p. 34a; and Scranton, p. 94)

Also as a result of discussions in *The Science of the Dogon,* comparisons between the Dogon myths and Egyptian language reveal the meanings of other key Egyptian glyphs. Dogon cosmology explicitly defines matter as the product of woven threads. The themes of weaving and threads recur throughout Dogon mythology but apply most specifically to the concept of matter as it is woven from vibrating threads by the spider Dada, whose name means "mother." These threads as they are described by the Dogon myths correspond closely to strings in string theory. *The Science of the Dogon* demonstrates that when we examine hieroglyphic counterparts to the Dogon keyword *thread,* we are led to the Egyptian word *ntt,* which is written two different ways.[4]

Two forms of the Egyptian word for thread (ntt)

The final glyphs in these two Egyptian words, ɣ and ╳, take the precise form of two diagrams from string theory that depict the two simple ways that strings are said to intersect. Likewise, when we follow Egyptian references for the concept of weaving, we soon arrive at the name of the Egyptian mother goddess Neith (Net), whose traditional role in Egyptian mythology was defined as the weaver of matter. Her name is written as follows:

Name of the Goddess Neith (Net)

Again, the final glyph in the name of the goddess Neith, ⊐⊂⊃⊏, is the very image of a string theory diagram that depicts a complex interaction of strings. And so we find that the Egyptian words used to express the concepts of weaving and threads center on recognizable scientific symbols—the very same symbols that are used by string theorists to represent a similar concept.

MEANINGS OF EGYPTIAN GLYPHS
***From* The Science of the Dogon**

Glyph	Meaning	Page Reference
ɣ	Simple string intersection	p. 99
╳	Simple string intersection	p. 99
⊐⊂⊃⊏	Complex string intersection	p. 100

We come away from *The Science of the Dogon* knowing likely symbolic meanings for thirty Egyptian glyphs—a number large enough to enable us to begin reading ideographic definitions for specific Egyptian hieroglyphic words. Each of these thirty glyphs has been explicitly defined, either by entries in Budge's *An Egyptian Hieroglyphic Dictionary,* by specific descriptions and drawings documented in *The Pale Fox,* or by the precise correspondence of the glyph with a scientific symbol. In our examples so far, we have successfully read a few Egyptian words and interpreted them in the way Budge says they originally could be read—by substituting defined ideographic concepts for each glyph in the word and then simply reading out a literal meaning. So with this set of tools, definitions, and examples in hand, we now are prepared to begin interpreting Egyptian hieroglyphic words.

FIVE

DEFINING EGYPTIAN GLYPHS

Up to this point in our study, we have identified five different sources from which we can draw information about the meanings of individual Egyptian glyphs:

- Sir E. A. Wallis Budge's List of Hieroglyphic Characters from his *An Egyptian Hieroglyphic Dictionary*
- The single-glyph and double-glyph word entries found in the body of Budge's dictionary
- Explicit definitions found in the descriptions of Dogon mythology
- Dogon cosmological drawings that match symbols of science
- Dogon keywords and their relationship to corresponding Egyptian words

Now we are about to add an important new sixth source to the preceding five, a source that has not been recognized in any previous study but that might well prove to be the most significant of the six—the Egyptian language itself. The clues that led us to this new source are drawn from an example in *The Science of the Dogon* and involve an Egyptian word relating to threads—pronounced "nu-t"—which means "to tie, to bind together."[1]

TO TIE, TO BIND TOGETHER (NU)

to tie, to bind together		(See Budge, p. 351a)

The glyphs used to write the word *nu* conform generally to what we would expect based on our knowledge of Dogon and Egyptian symbols and string theory. They include the wave glyph, which can refer to waves in quantum theory or to the concept of weaving, the Egyptian adze glyph ( ), the clay pot glyph ( ), which we have tentatively identified as representing a particle, the spiraling coil glyph ( ), which represents the mythical Dogon thread, and the looped string glyph ( ). When it comes to the structure of the word *nu* itself, we are familiar with both the wave glyph ( ) and the particle glyph ( ), but the glyph that stands between the wave and the particle glyphs is one whose mythological meaning we have not yet defined—the Egyptian adze glyph ( ). Budge tells us that, in common usage, this glyph represents the blade of an adze and can mean "to form" or "to shape." The adze is an ancient woodworking tool, similar to a wood plane, that is used to shave or shape wood. The positioning of this glyph within the word *nu* suggests that it might somehow represent the process by which an unformed wave of matter comes to be shaped into a particle.

We can find many ancient and modern references to the Egyptian adze, but few tell us much more than we already know—that it is an ancient tool used to shape or form wood. If we want to learn more about the mythological symbolism of the adze, we find that our only real source of information is the Egyptian language itself. This realization eventually leads us to examine the Egyptian words for *adze* in the hope of discovering some clue as to its larger meaning.

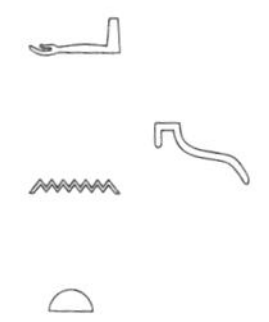

An Egyptian Word for Adze (an-t)

One Egyptian word for adze is *an-t*. Its glyphs include the bent arm glyph ( ), which we believe implies an action or force, the wavy line of water glyph ( ), and the hemisphere glyph ( ), which we believe symbolizes the concept of mass or matter. The word also includes the adze glyph itself, which is somewhat surprising, because if the Egyptians had been inclined to use a picture of the adze blade to represent the word "adze," they could have done so using just that one ideographic glyph. So what is it that the adze glyph really tells us in combination with these other glyphs? Is it simply a determinative used to establish context, as Budge suggests, or does it serve some other purpose? Upon reflection, we realize that if we set the adze glyph aside—as it is already physically set aside in the actual hieroglyphic word—and simply apply our tentative ideographic definitions in place of the other three glyphs, the meaning we read out is almost precisely the one we suggested for the adze glyph during our discussion of the word *nu* above.

DEFINITION OF THE ADZE GLYPH (AN-T)

Adze		The act or force by which waves attain mass , followed by the adze glyph . (See Budge, p. 123b)

If this ideographic interpretation is a correct one, then we quickly realize that the adze glyph does not appear among the glyphs of the word in the role of a determinative, but rather it appears as the final glyph of the word because, ideographically, the preceding glyphs of the word appear to specifically define its meaning. If that is the case, then there may well be an unnoticed convention at work within the Egyptian hieroglyphic language—a convention of defining words whose purpose may be to explicitly define the meanings of individual glyphs.

With that thought in mind, if we now take a second look at the Egyptian word *nu,* we see its component glyphs in a new light. There is now cause to suspect that the purpose of this word, which means "to tie, to bind together," might actually be to define the last glyph in the word, the looped string glyph ( ), whose shape is documented in string theory as being that of one of the simple intersections of a string. If we follow the example of the word *adze* above and apply the ideographic concepts of each glyph to the word but exclude the final glyph—the glyph that seems to be defined by the word—we see that the result is a very appropriate scientific definition of the looped string intersection.

DEFINITION OF THE LOOPED STRING INTERSECTION GLYPH (NU)

To tie, to bind together		Waves form and/or shape particles from coiled threads , followed by the looped string intersection. (See Budge, p. 351a)

With these two initial examples, we can infer from their ideographic forms a symbolic role for the Egyptian adze glyph in the process of the tying and binding of threads and particles. Yet, although these examples may be suggestive, we must stop for a moment and ask ourselves whether there is any real evidence in the Egyptian language that would justify the association of what seems (in common usage) to be a simple woodworking tool (the adze) with rudimentary materials of weaving (cords and threads). We can answer this question with a direct and unconditional yes, simply by examining a second Egyptian word for adze, this one pronounced "nu"—the very same pronunciation as the word discussed above meaning "to tie, to bind together."

DEFINITION OF THE ADZE GLYPH (NU)

Adze		Forms/shapes particles from coiled threads. (See Budge, p. 352a)

The word *nu,* meaning "adze," is written using the same three central glyphs as the word *nu,* meaning "to tie, to bind together," and is pronounced in the same way. In fact, we should note that in Budge's word entry, one spelling of the word *nu* is written using a single ideographic glyph—the adze glyph ( ) itself—just as we surmised it could and should be. Three common aspects of these two words lead us to associate the Egyptian *adze* with the word for tied and bound threads: the three central glyphs that form each word, their common pronunciations, and their common ideographic meanings.

The success of these initial examples suggest that our observation may be a correct one—that the Egyptian hieroglyphic language may well employ a system of defining words as a way of explicitly describing the meanings of its own glyphs. One way to test this proposition further would be to use it to try to validate the definitions of other Egyptian glyphs whose meanings we believe we already know. Based on the pattern presented to us by these first examples, we can establish the following tentative criteria for identifying an Egyptian defining word: the common-usage meaning of the word should express the ideographic concept depicted by the glyph, and the written form of the word should display the defined glyph as the final glyph of the word.

In the previous example, we successfully determined the meaning of the looped string intersection; now let us test our proposed convention using a related glyph that defines the complex string intersection, which is the scientific process that might truly be said to weave matter. Egyptian mythology is quite clear in its assignment of the role of weaver of matter to the goddess Neith. She was the great mother goddess who was said to have given birth to all the other Egyptian gods. As our criteria require, we notice in the following example both that the word expresses the ideographic concept we believe is symbolized by the glyph—that of weaving matter—and that the final glyph in the hieroglyphic name of Neith (Net) is the one that depicts the complex intersection of strings. When we apply ideographic meanings in place of the glyphs of the word *Net,* they read as follows:

DEFINITION OF COMPLEX STRING INTERSECTION GLYPH (NET)

Name of Neith		Weaves matter , followed by the complex string intersection and the goddess glyph determinative . (See Budge, p. 399b)

This reading of the name of Neith depends upon the symbolic assignment of the concept of mass or matter to the hemisphere glyph . Although the interpretation of "mass or matter" seems justified based on references from Dogon cosmology,[2] we need to ask ourselves whether this same symbolism can be justified in terms of the Egyptian language. Once again, we find that the reference can be affirmed if we look to the Egyptian word *pau-t,* which Budge defines as meaning "stuff, matter, substance, the matter or material of which anything is made."[3] It relates to the Egyptian word *pau,* which corresponds to the Dogon po, the mythological counterpart to the atom.[4] The ideographic sense of this word depends on the symbolic meaning of the flying goose glyph , which is pronounced "pa" (see the word *pa,* meaning "to be, to exist")[5] and the glyph, which Budge interprets as symbolizing primeval time.[6] The word *pau-t* is written as follows:

DEFINITION OF HEMISPHERE GLYPH (PAU-T)

Mass, matter		Existence primeval (mass or matter), followed by the hemisphere glyph . (See Budge, p. 230b, "stuff, matter, substance, material of which anything is made"—next to last spelling)

Although in the context of his hieroglyphic dictionary entry Budge does not explain the assignment of "primeval time" as the meaning of the glyph, both the shape and definition of the glyph make sense in terms of key corresponding references in *The Pale Fox*. These references describe events that transpire at the time of the formation of matter, which we interpret as primeval time,[7] and involve tearing off a square piece of a placenta. And so both the ideographic form of the glyph and Budge's definition make sense in terms of what we know of Dogon cosmology. We can see from each of the above examples that our proposed defining word convention produces appropriate definitions for the role of the goddess Neith in Egyptian mythology, for the complex string intersection glyph , and for each of the supporting glyphs required to interpret these words ideographically.

Continuing along this same line of inquiry, we now turn to another Egyptian word for the concept of a thread—the word *nt-t*. We see from the example below that the concept of an Egyptian defining word is upheld, but this time in a slightly modified form. Based on its configuration, this word might be more aptly described as an enumerating definition. It employs the conventions of an Egyptian defining word to describe a set of related Egyptian glyphs—in this case, the two simple string intersection glyphs and . The convention still provides for the definition of the final glyphs of the word based on the ideographic meanings of the leading glyphs, however in this case it is two glyphs—not one—that are being defined.

DEFINITION OF THE SIMPLE STRING INTERSECTION GLYPHS (NT-T)

Thread		Weaves matter in two ways, followed by the and string intersection glyphs. (See Budge, p. 399b)

We know that Egyptian mythology—like that of the Dogon and many other ancient cultures—refers to the notion of the weaving of matter. For example, in Egyptian mythology, the goddess Neith weaves matter with her shuttle, and in Dogon mythology, the spider Dada is credited with weaving matter. There is a kind of cross-cultural mythological metaphor at work that seems to equate the concept of weaving with the concept of matter. The question we should ask is whether there is any evidence of a relationship in the Egyptian language between the concept of weaving and the glyph shapes we associate with the formation of matter based on references from Dogon cosmology—in particular, the hemisphere glyph and the coiled thread glyph.

When we examine Egyptian words that mean "to weave," we soon come to the word *skhet.*[8] (A word with the same pronunciation can also mean "to erect a shelter made of leaves and branches," a meaning that calls to mind the *sukkah* of Judaism.) This word takes the form of a defining word for the coiled thread glyph, which is the Dogon mythological equivalent of a string after it has passed through the seven wrapped-up dimensions of the Calabi-Yau space. It is written using the curved staff glyph, which we interpret ideographically as meaning "to bend" (see discussion below); the glyph, which Budge defines as "a sieve"[9] (see discussion of the sieve glyph in the next chapter); the hemisphere glyph; and the coiled thread glyph. Together, we interpret these glyphs ideographically to mean the following:

DEFINITION OF THE COILED THREAD GLYPH (SKHET)

To weave	[glyphs]	The bending [glyph] sieve [glyph] of mass [glyph], followed by the coiled thread glyph [glyph]. (See Budge, pp. 694b, 695a)

Although the image of the coiled thread ([glyph]) is an important one in Dogon cosmology, it is not one that is explicitly conceptualized in string theory. It does, however, present us with a possible reconciling link to torsion theory, whose tenets depend upon a spiral-shaped pleating or folding of space-time. This spiral is interpreted by some as a kind of vortex within the hyperspace or subspace that structures space-time. Budge actually defines an Egyptian word *pekhar-pekhar,* meaning "vortex."[10] This word is written using the spiraling coil glyph and reads ideographically as follows:

THE CONCEPT OF A VORTEX (PEKHAR-PEKHAR)

Vortex	[glyphs]	The circuit [glyph] of the spiraling coil's [glyphs] flow [glyph]. (See Budge, p. 247a)

Torsion theory, like Dogon cosmology and string theory, considers matter to be woven from threads and considers particles to be the result of what are essentially knots in these primordial threads. This same concept may be expressed ideographically by the Egyptian word *atennu,* meaning "knots."[11] This word comes from the phonetic root "aten," which Budge defines as meaning "to bind, to tie." The word *atennu* is written as follows:

DEFINITION OF THE KNOT GLYPH (ATENNU)

Knots		That by which mass is woven into particles from the coiled thread, followed by the knot glyph, and the number 3 determinative (threads intersect/are tied in three ways). (See Budge, p. 99a)

The concept of bending in relation to the formation of mass or matter is an important one in string theory. The seven wrapped-up dimensions of each Calabi-Yau space cause a vibrating string to bend in new ways, and it is through this process of bending that the string is thought to attain mass. We can relate this process to the curved staff glyph by way of two dictionary entries for the Egyptian word *au-t,*[12] one of which Budge defines as meaning "stick with a curved end," the other as "staff, crook, scepter." The glyphs of the word *au-t* include the chicken, or quail, glyph, which can mean "place" (see the defining word entry *b*, meaning "abode, place"[13]), or can imply "growth" (see the Egyptian word *rutu,* meaning "growing plants"[14]).

DEFINITION OF THE CURVED STAFF GLYPH (AU-T)

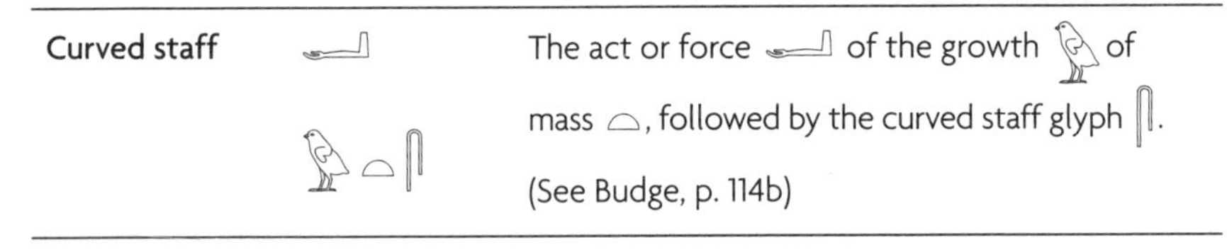

Curved staff		The act or force of the growth of mass, followed by the curved staff glyph. (See Budge, p. 114b)

The implication of these initial examples—if they are borne out by other Egyptian words—is a potentially significant one; it strongly suggests that we can use built-in conventions of the Egyptian hieroglyphic language itself to validate the likely ideographic meanings of individual Egyptian glyphs. These preceding definitions corroborate what we have already been explicitly told about many of these shapes in Dogon mythology and reaffirm the specific correlations we have made between those shapes and symbols of science. With this new convention in hand, if we now find ourselves in doubt about the larger symbolic meaning of a given Egyptian glyph, our first impulse might be to locate an Egyptian defining word and consider what that word may tell us about the ideographic meaning of the glyph.

MEANINGS OF EGYPTIAN GLYPHS
Interpreted from Defining Words

Glyph	Meaning	Term and Budge Page Reference
	The act or force by which waves attain mass	*an-t,* p. 123b
	Waves form/shape particles from coiled threads; weaves matter	*nu,* p. 351a; and *nt-t,* p. 399b
	Weaves matter	*Net,* p. 399b
	Existence primeval (mass, matter)	*pau-t,* p. 230b
	Weaves matter	*nt-t,* p. 399b
	The bending sieve of mass (coiled thread)	*skhet,* pp. 694b, 695a
	Place, to grow	*b,* p.197a; and *rutu,* p. 421b
	Bending force of the growth of mass	*au-t,* p. 114b

From *An Egyptian Hieroglyphic Dictionary.*

In each of the above cases, we can see that our definition of an Egyptian defining word proves itself. We find that when we substitute ideographic meanings for the glyphs of each sample word, we are able to read out a meaningful definition for each defined glyph. The preceding examples also provide us with another set of defined Egyptian glyphs, which we can include in our growing list. We realize that because Budge's *An Egyptian Hieroglyphic Dictionary* consists of two volumes of Egyptian words and his List of Hieroglyphic Characters includes more than fourteen hundred Egyptian glyphs, this list is one that is likely to expand greatly as our experience with the Egyptian hieroglyphic language grows.

SIX

EGYPTIAN CONCEPTS OF ASTROPHYSICS

At this point in our study, given the many consistent correlations that have been shown to exist between symbols of Dogon and Egyptian mythology and modern science, it would not be unreasonable to suggest that the symbols of Dogon and Egyptian myth and language were intentionally designed to describe the underlying structure of matter. This would imply the following assumptions:

- Someone at or around 3400 BCE was in command of many intimate details of science and deliberately chose to encode them in the structures of myth and language.
- They knew the correct shapes of various components of matter as defined in atomic theory, quantum theory, and string theory.
- They knew the proper hierarchy of these constituent elements of matter within the larger structure.
- They were aware of the relationships among the components and forces of matter and how each of these components behaves.

In other words, they understood *astrophysics*.

Yet the science of astrophysics involves more than just the rote recitation of the component elements that comprise matter. Any correct understanding of these components rests on basic concepts and principles that underlie the processes by which they are formed—concepts that for modern cultures only came to be understood slowly, over long periods of time, and as a result of the successive contemplations of some of the most insightful minds in history. Knowing this, we must assume that if some ancient authorities knew about the each of the various subcomponents of matter, then they also understood these underlying concepts. If there was an intention to deliberately encode details of astrophysics into ancient mythology and language, then it is reasonable to think that we might find information about these concepts reflected in the Egyptian hieroglyphic language.

We spoke previously about Albert Einstein's theory of relativity and his formula $E = mc^2$, which defines a three-way relationship between mass or acceleration, time, and the speed of light. One consequence of this formula is that as a body is accelerated (or its mass is increased), time must slow down. This relationship between mass and time is governed by the speed of light, which according to Einstein must be measured at a constant 186,000 miles per second by all observers, regardless of their own motion. From this perspective, it is appropriate to view light as the limiting factor of mass and time.

If the theoretical designer of the Egyptian language had full knowledge of the concept of light, then we would expect the Egyptian hieroglyphs to somehow reflect that knowledge. The Egyptian word for light is *aakhu,* a word whose pronunciation corresponds favorably with the Dogon word *Ogo,* the name of an allegorical character from Dogon mythology who seems to symbolize light. There is also a related Egyptian deity named Aakhu who, according to Budge, was the Egyptian god of light. We can interpret *aakhu* as a defining word for the light glyph because—as our criteria require—it describes the ideographic concept depicted by the glyph and includes the light glyph as its final character. The word *aakhu* includes another Egyptian glyph that we have mentioned but not explored—the glyph, which Budge interprets as a sieve, although his dictionary defines no Egyptian word for sieve that includes it.

The concept of the sieve is an important one in the Dogon tradition and one that *Dictionnaire Dogon* suggests can be understood by way of metaphor. In both the Dogon language and the Egyptian hieroglyphic language, the phonetic value "nu" implies the concept of water. However, for the Dogon, the word *nu* can also carry the meaning of "bean." In *Dictionnaire Dogon,* Genevieve Calame-Griaule tells us that the Dogon preserve beans by covering them with sand. To use the beans later, the Dogon employ a bean sieve to separate them from the sand. This process is considered a metaphor for the act of divination, which the Dogon define as the process of separating truth from error.[1] However, divination is also a process associated with water, and so the act of separating the seedlike beans from the waterlike sand might also be seen as a metaphor for the emergence of particles from waves. When the Dogon priests initiate apprentice diviners, they always use a traditional ancient round sieve that is symbolic of the round divination table that Germaine Dieterlen depicts in *The Pale Fox.*[2]

An alternate, if somewhat unorthodox, interpretation for the Egyptian sieve glyph shape, based on the Dogon pronunciation of "nu" and its use in an ideographic context, would be that it depicts a pool of water and carries one of three meanings: "source" (as a lake can be considered the source of a river), "limit" (as the bank of a lake can be considered the limit of the water it contains), or "effect, result, or product" (as a collected pool of water is the effect or product of accumulated rainfall). In support of the last of these symbolic meanings, the word *khet,* written 𓐍𓏏, means "products of."[3] The "pool of water" interpretation itself is supported by the Egyptian word *kh[i],* written 𓐍𓀠, a word whose ideographic characters directly convey its meaning as defined by Budge—that of "high or rising water."[4] Such words may constitute another class of Egyptian defining words that I call contextual definitions.

Ideographically, the meaning "pool of water" becomes apparent in a series of Egyptian words. The word *kheb* [5] means "marsh" and reads ideographically as "pooled water, place of plants"—a precise definition of a marsh. The word *khebs* ,[6] meaning "a diving bird," reads ideographically as "pool of water, place of a diving bird." For the word *khefi-t,*[7] meaning "quay, shore, bank, landing-stage," in four out of five spellings, the sieve glyph offers the only likely ideographic reference to water, and the word itself carries the precise meaning we interpret for the glyph. The word *Kha*[8] refers to "a lake in the Tuat," and Budge's second spelling for the word depicts the lake ideographically using the same glyph. Rising water, we know, acts as nature's sieve, as was amply demonstrated by the annual flooding of the Nile in Egypt, an event that deposited rich soil used for planting after the waters receded. Based on these kinds of references, we see that the alternate "pool of water" interpretation for this glyph is both sensible and supported. Either interpretation, that of a sieve or that of a pool of water, provides an appropriate metaphor for the concepts of source, limit, and product or result. When we apply our ideographic meanings to each of the glyphs in the defining word *aakhu,* they read as follows:

DEFINITION OF THE LIGHT GLYPH (AAKHU)

Light		That which comes to be the limit , followed by the light glyph . (See Budge, p. 23a)

Dieterlen devotes an entire chapter of *The Pale Fox* to the character of *Ogo,* one of the principal players in a mythological episode in which the universe is formed. According to Dogon mythology, at the time of the formation of the universe Ogo became impatient for Amma's creation to unfold. Ogo thought that he could create a universe as perfect as Amma's, and so he stole a square piece of her placenta, which Ogo believed to contain his twin sister. However, Amma had foreseen Ogo's actions and placed the spiritual principle of the twin sister beyond Ogo's reach. From that time on, as a consequence of his own impulsiveness, all of Ogo's future efforts would be devoted to trying to regain his lost twin.

We can interpret this episode of Dogon mythology as a parable about astrophysics. In the analogy, if Ogo represents light, as the corresponding Egyptian word *aakhu* confirms, then Ogo's twin sister would represent time. Likewise, the square that Ogo stole from Amma's placenta would symbolize space, the ideographic concept that we assign to the Egyptian square glyph □ based on its appearance in words relating to the concept of spaciousness.[9] The episode about Ogo as it is described in *The Pale Fox* specifically states that these events happened outside of space and time. In fact, according to Dieterlen, it was these very acts by Ogo that "prefigured" the attributes of dimension, space, and time in the universe. If we apply the symbolic meanings of "light," "time," and "space" to the characters of the Dogon myth, then we can see that the eternal chasm that came to exist between Ogo and his twin sister at the time of the formation of the universe expresses the essential nature of the relationship between light and time as it is defined in modern astrophysics.

The next key concept of astrophysics to be examined in the context of the Egyptian language is that of gravity, and the most likely place to begin our comparisons is with the theories of Einstein. His view of gravity is based on a concept he proposed but could not prove. He suggested that we think of the universe as being comprised of an invisible fabric called space-time. This proposed fabric would be the scientific equivalent of the classical Greek concept of the aether—a theoretical fluidic essence thought to fill the space between stars and planets. According to Einstein, the fabric of space-time bends and warps in response to the various bodies of mass that it surrounds. The bending or warping of this fabric is responsible for two main effects that we observe: it causes unsupported objects to fall to Earth, and it causes astronomical bodies like Earth to orbit around other bodies like the sun. When we examine Egyptian words that express the concept "to fall" or "to fall down," we see that they often are expressed in terms of the mouth glyph ⊂⊃, a symbol whose very image seems to convey the ideographic concept of bending, warping, or swelling.

DEFINITION OF THE FALLING MAN GLYPH (KHER)

To fall down	⊜ ⊂⊃ [falling man glyph]	Result/product/effect ⊜ of the bending/warping force ⊂⊃ followed by the falling man glyph [falling man glyph] (effect of gravity). (See Budge, p. 560b; and the word *sher,* p. 749b)

We can confirm our suspicion that the mouth glyph symbolizes the concept of bending, warping, or swelling by looking to the Egyptian word *ref,* which means "to swell up" (see Budge, p. 423b). It is written with the glyphs ⊂⊃ [horned viper glyph] [standing figure with raised hands glyph], which we interpret ideographically as meaning "warping or swelling transmitted upward." As we observed with the Egyptian word for rising water, the standing figure with raised hands glyph [standing figure with raised hands glyph] seems to convey the ideographic concept of up, while the horned viper glyph [horned viper glyph] appears to reflect the concept of "transmission." By that logic, the notion of swelling would be represented by the mouth glyph ⊂⊃—a figure that can be seen as thin on both ends and swollen in the middle.

The second major effect of gravity that we observe is that of an orbit—the tendency of astronomical bodies to circle around other bodies. Budge defines two Egyptian words meaning "to encircle, to orbit," the words *shenu*[10] and *qet,*[11] meaning "circle, orbit." These words qualify under our criteria as defining words for the Egyptian circle glyph ○, which we can interpret as meaning "the path an orbiting body makes due to gravity." The circle glyph provides a clue for understanding other words whose meanings are based on the concept of an orbit. For instance, the word *year* expresses a fundamental unit of time, the time that it takes for Earth to orbit the sun. To understand the ideographic meaning of this word, we must first be aware of a character that Budge explicitly identifies in his List of Hieroglyphic Characters as the Egyptian time glyph 𓆳.[12] He does not venture an opinion as to which object might be depicted by the glyph—modern Egyptologists interpret it as representing a sprouting plant. The Egyptian word *renp-t,* meaning "year," is written as follows:

DEFINITION OF A YEAR (RENP-T)

Year	𓆳 ☉	The time 𓆳 of Earth's orbit around the sun ☉. (See Budge, p. 427b)

In this word, the glyph that we would interpret as depicting an orbit is the Egyptian sun glyph ☉. Our reading suggests that the sun glyph comprises two figures—the circle glyph ○, which we interpret as the circular path of an orbiting body, and a central dot. Of course, when we speak in terms of the concept of a year, the orbiting body we are referring to is Earth. By this logic, the central dot would be the body around which Earth orbits, the very body that is symbolized in the Egyptian language by the compound glyph—the sun.

The idea that the concept of gravity might be portrayed by the image of an orbit is one that is entirely in keeping with the mind-set of Dogon cosmology as we understand it, wherein the mythological counterpart to an electron (the sene seed) is illustrated with a drawing that closely resembles an electron orbit [13] and the concept of the vibration of a thread is explained by a drawing that resembles one of the typical vibratory patterns of a string in string theory ⊗.[14] These images represent the most obvious effect of the force or component of matter being described and take the form of familiar scientific diagrams, of the sort that would be immediately recognizable to an astrophysicist.

Other Egyptian language references help us to validate our interpretation of the Egyptian sun glyph and confirm that it can be properly interpreted as depicting an orbit. Initial validation comes from Budge's entry for the Egyptian word *abt,* which means "month."[15] This word is written with three Egyptian characters: the glyph, which Budge defines to mean "moon" (see Budge, p. cxxv), the hand glyph , which we interpret to mean "to give or to make," and the sun glyph , which we believe depicts an orbit. Ideographically, the word *abt* reads as follows:

DEFINITION OF A MONTH (ABT)

Month		The moon makes an orbit . (See Budge, p. 40b)

Our ideographic reading of the word *abt* reveals yet another Egyptian word that appears to define its own scientific meaning, and once again the meaning depends upon our initial interpretation of the sun glyph as depicting an orbit.

The sun glyph can again be seen to symbolize Earth's orbit around the sun in the Egyptian word *ses,* which means "seasons."[16] The clarity of the definition provided by the word lies with the knowledge that the ancient Egyptians observed only three seasons: a planting season, a harvest season, and a rainy season. Also included within this word is a different kind of trailing determinative than we have previously seen—the number three ( ) written in Egyptian glyphs, which traditionally is interpreted as identifying a word as being plural.

DEFINITION OF SEASONS (SES)

Seasons	∩∩∩ ⊙ ı ı ı	Three bends/arcs ∩∩∩ of Earth's orbit around the sun ⊙ followed by the number 3 determinative ı ı ı. (See Budge, p. 696ab)

Based on our own emerging standards of interpretation, if the Egyptian sun glyph were to depict an orbit and represent an effect of the bending or warping force of gravity, then we would expect to find an Egyptian defining word for the sun glyph that supports that view. Appropriately, we find just such an example in the spelling of the word *Ra,* a traditional Egyptian name for the sun and the name of the sun god.

DEFINITION OF THE SUN GLYPH (RA)

Sun glyph	⊂⊃ (mouth) / arm / ⊙	The bending/warping ⊂⊃ force (arm), followed by the sun glyph ⊙—the image of an orbit. (See Budge, p. 417b)

We can further support our interpretation of the Egyptian word *Ra* as representing the bending/warping force of gravity by looking again to the Dogon language. In *Dictionnaire Dogon,* Calame-Griaule defines the word *se:re* as meaning "to warp." Because the Dogon prefix *se-* means "to have" or "to possess," in the sense of possessing a quality, then a more precise translation of the word *se:re* might be "to possess the quality of warping." By this interpretation, the suffix *-re* is a likely Dogon counterpart to the Egyptian word *Ra,* and like the mouth glyph (⊂⊃), which in Budge's view is pronounced "r" or "ra," it would mean "to warp." One interesting feature of the Egyptian hieroglyphic language is that the sun glyph is used in some contexts to represent the sun and in others to represent a unit or period of time. In the examples above, we can see clear justification for both symbolic meanings.

The question now arises as to what scientific mechanism may be described in the Egyptian language to explain the cause of this gravitational effect—the tendency of one astronomical body to orbit another, an effect that is at the heart of the definition of some of our most fundamental units of time. Again, our experience with Egyptian hieroglyphs suggests that we should expect to find an example of this process in the use of an Egyptian defining word for the sun glyph. If we look to hieroglyphic words that express concepts of time and seasons, we soon arrive at the Egyptian word *aether,* the very word used by ancient Greek philosophers who may have studied with Egyptian priests to describe a cosmic quintessence or fabric. Budge interprets the word *aether* to mean "time or season," a definition that our examples have shown to be directly related to the concept of an orbit. Because this word both describes a traditional meaning of the sun glyph and exhibits that glyph as the final glyph in the word, it can be seen as an alternate defining word for the Egyptian sun glyph ☉. In this case, each of the characters that comprise the word *aether* has an established ideographic meaning in our lexicon of interpreted glyphs, so we are able to simply substitute these meanings and read out a likely definition for the word.

THE BENDING FORCE CAUSES AN ORBIT (AETHER)*

Definition of gravity	⌓ ⬭ 𓎛 ☉	Mass ⌓ bends/warps ⬭ time 𓎛, followed by sun glyph ☉. (See Budge, p. 101b)

*Budge interprets this word to mean "time or season."

Based on the ideographic interpretation of the word *aether,* the processes that produce an orbit conform generally to what Einstein proposes in his theory—an orbit is defined as a by-product of warping or bending that is induced by mass. However, the Egyptian definition differs in one significant respect from that proposed by Einstein; for the Egyptians, it is not space-time, but rather time itself that is bent by mass. Unlike Einstein, the definition drawn from the Egyptian language does not postulate some new conceptual fabric to convey the force of gravity; instead, it suggests that the force of gravity is something that is transmitted by the fabric of time itself, that gravity is what happens when mass bends time. It is interesting to note that this concept is explicitly supported by one of Einstein's own postulates, which states that the laws of physics in an accelerating frame are indistinguishable from those in a gravitating frame. Because Einstein himself tells us that it is time that is bent by acceleration, then for the effects of gravity to indistinguishably match those of acceleration, must not the effects of gravity also derive from the bending of time?

From this perspective, the aetherlike fabric that Einstein calls *space-time,* the fabric that could be said to bind space itself and transmit the force of gravity, would, in essence, simply be *time.* We can see supporting evidence for this perspective in the Egyptian hieroglyphic language in the form of another defining word meaning "time or period," which Budge pronounces "sab."[17] To understand this definition, we must first identify yet another Egyptian glyph, an alternate character that Budge also defines as meaning "time" (see Budge, p. cxlv). Based on that definition, the word *sab* reads as follows:

DEFINITION OF THE TIME GLYPH (SAB)*

Time	□	Binds space □, followed by the time glyph. (See Budge, p. 588b)

*Budge interprets this word to mean "time, period."

In this light, we are now prepared to understand another Egyptian word that defines the concept of time. This word is *atru,* and it meets our criteria as a defining word for the time glyph. Except for the time glyph itself, all characters used to write the word *atru* fall within our growing lexicon of defined glyphs, so we can interpret a meaning for the word based solely on established definitions. When we apply these ideographic meanings to the glyphs, we see that they express the same relationship of time to mass as is implied by Einstein's theory of relativity: time slows or bends as mass increases.

DEFINITION OF THE TIME GLYPH (ATRU)

Time		That which mass bends or warps , followed by the time glyph . (See Budge, p. 100a)

Moving now beyond notions of gravity and time, the next concept of astrophysics for us to consider would be that of acceleration. We recall that Einstein's theory of relativity is presented in the context of the acceleration of a body of mass. In the view of astrophysicists, the acceleration of a body is equivalent to an increase in its mass, and according to Einstein, either event causes the same slowing of the time frame for the affected body. The Egyptian word that means "accelerate" is *s-khakh*. This word includes four glyphs we have not encountered in previous examples. These glyphs are , which we can interpret based on its ideographic form to depict an increase in mass; , which conveys the concept of "by" (Budge says that it can also imply duality); , which is a cutting tool—it implies the mathematical concept of subtraction; and , a glyph that Budge interprets to mean "strength" (see Budge, p. cvii). Like the sun glyph, this last glyph can actually be seen as consisting of two figures—the bent arm glyph holding the sprouting plant glyph . We know that the bent arm glyph means "to bear, to carry, anything done with the arm" (see Budge, p. cvii), but it also can mean "piece."[18] We know from our example with the word *atru* above that the sprouting plant glyph means "time." So the broader implication is that the glyph refers to the strength of something or a piece of something and may relate to the concept of time. Based on that interpretation, we see that the Egyptian concept of acceleration is so similar to Einstein's that it could almost serve as a textbook description of his theory. Together, the glyphs of the word *s-khakh* read as follows:

DEFINITION OF ACCELERATION (S-KHAKH)

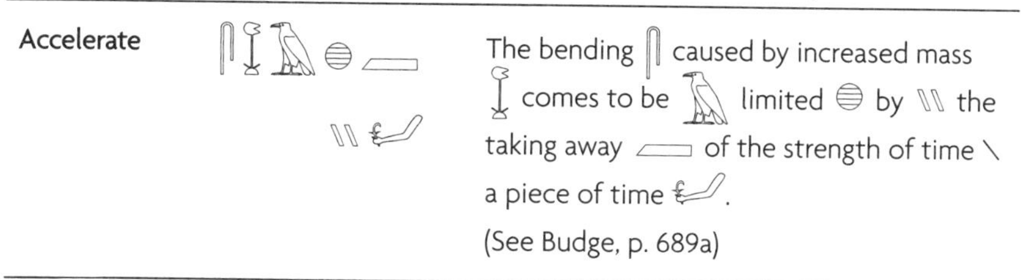

Accelerate	The bending caused by increased mass comes to be limited by the taking away of the strength of time \ a piece of time. (See Budge, p. 689a)

Last but not least, we now turn our gaze to examine various Egyptian references relating to the other quantum forces that are seen as companions to gravity. As we mentioned in a previous chapter, our ability to interpret these Egyptian references is aided by a discussion in *The Science of the Dogon* in which each of the four quantum forces is equated to one of four branches of the mythological Dogon sene na,[19] whose names in the Dogon language mean "to bring together," "bumpy," "that bows its head," and "stocky."

The second of the four quantum forces is the electromagnetic force, which causes an electron to orbit the nucleus of an atom. One modern-day scientific symbol for the electromagnetic force is a zigzag line ⩘, a figure that closely resembles the Egyptian wave glyph. Ideographically, this image is a good match for the second branch of the Dogon sene na, called *sene gommuzu,* whose name in the Dogon language means "bumpy." The electron itself is one of the likely scientific counterparts to the Dogon sene seed, a mythological building block of the atomlike po that is said to surround the po and cross in all directions to form a nest. One Egyptian word for nest is *au,* the phonetic root of the Egyptian word *aun,* which means "to open, to make to be open."[20] This is a word that we have already interpreted in the context of the electromagnetic force in chapter 4. It is written with what we will refer to as the nest glyph ⊹, a character whose image matches both the Dogon mythological drawing of the nest of the sene seed and the shape of a typical orbital pattern of an electron as it orbits the nucleus of an atom.

AN ELECTRON (AUN)

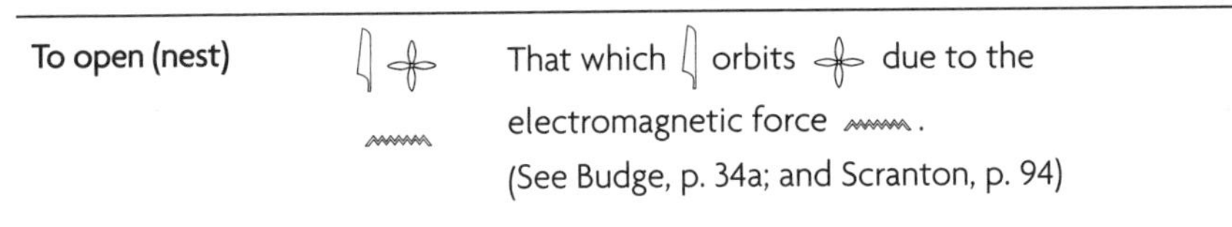

To open (nest)		That which orbits due to the electromagnetic force . (See Budge, p. 34a; and Scranton, p. 94)

We can reaffirm the meaning of the nest glyph by looking to Budge's entry for the glyph in his List of Hieroglyphic Characters.[21] He offers no explanation for the image depicted by the glyph but includes Egyptian glyphs as a reference to its phonetic value of "un." One of Budge's entries is written using the chicken or quail glyph, which we previously suggested can mean "place" or "growth." Ideographically, Budge's entry reads as follows:

DEFINITION OF NEST GLYPH ( )

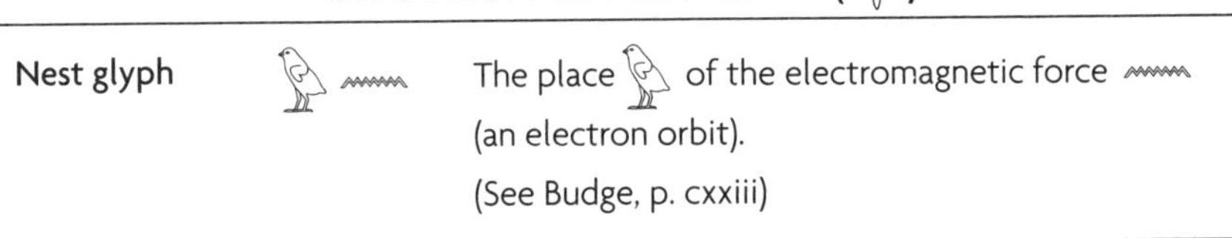

Nest glyph		The place of the electromagnetic force (an electron orbit). (See Budge, p. cxxiii)

A third quantum force—the weak nuclear force—corresponds to a third branch of the Dogon sene na, known as *sene urio,* whose name in the Dogon language means "that bows its head." The weak nuclear force causes quarks to be bound together into larger particles such as protons and neutrons. We can relate this Dogon phrase to the Egyptian language based on the word *sen,* which according to Budge means "to bow or pay homage."[22] Together, the glyphs of the word *sen* read as follows:

DEFINITION OF THE WEAK NUCLEAR FORCE GLYPH (SEN)

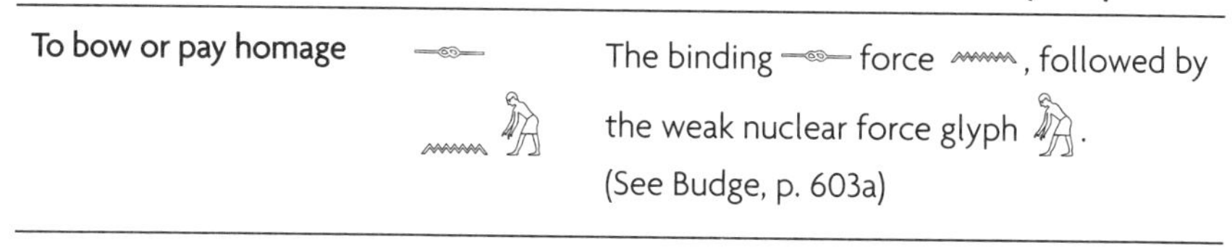

To bow or pay homage		The binding force, followed by the weak nuclear force glyph. (See Budge, p. 603a)

The last of the four quantum forces is the strong nuclear force, which tightly binds the components of an atom together. It corresponds to the fourth branch of the sene na, called *sene benu,* whose name in the Dogon language means "stocky." In Dogon mythological terms, this is the force that "draws the atom." Its counterpart in the Egyptian language is the word *sennu,* which Budge defines to mean "image." This is the same keyword that the Dogon use to define the mythological atom when its components have been bound together by the strong nuclear force. The word *sennu* includes two wave glyphs together, which might imply the concept of a *force of forces*—an appropriate description when applied to the strong nuclear force.

DEFINITION OF THE STRONG NUCLEAR FORCE GLYPH (SENNU)

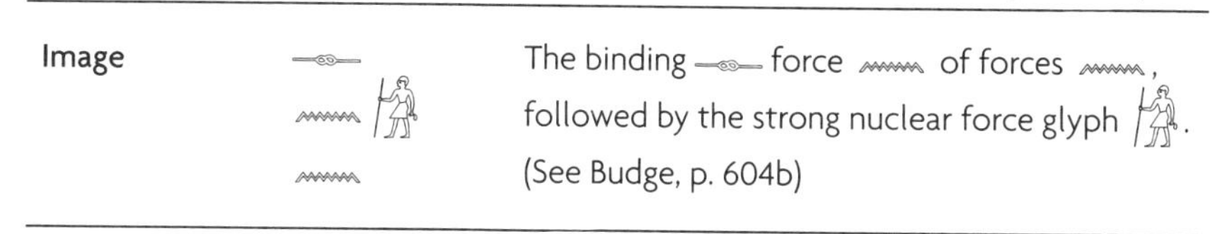

Image		The binding force of forces, followed by the strong nuclear force glyph. (See Budge, p. 604b)

It is interesting to note that the Egyptian hieroglyphic language uses two glyphs with related ideographic images, both involving figures of a person, to define what appear to be two related quantum forces—the weak nuclear force and the strong nuclear force. The weak nuclear force is defined, in accordance with the Dogon term meaning "that bows its head," by the image of a bowing man. Similarly, the strong nuclear force is represented, also in accordance with its corresponding Dogon term that means "stocky," by the image of an upright, robust man.

This symbolic approach is in agreement with the use of the falling man glyph to represent the effect of gravity.

In previous word examples (here and in *The Science of the Dogon*) we have defined four separate glyphs to represent various kinds of forces of astrophysics: the bent arm glyph , which implies "a force" in the generic sense; the mouth glyph , which represents "a bending or warping force"; the drawing board/loom glyph , which represents "a drawing/weaving force"; and the wavy line of water , which can imply "a wave force." The meaning of each of these glyphs was derived from corresponding symbolic themes in Dogon cosmology and from its appearance in Egyptian words and phrases that bear a relationship to concepts of astrophysics. However, it is interesting that Budge himself defines a relationship among these same four glyphs when he defines the phonetic value of the bent arm glyph in his List of Hieroglyphic Characters.

DEFINITION OF THE BENT ARM GLYPH ("TO BEAR, CARRY, SET IN POSITION, ANYTHING DONE WITH THE ARM" [ ])

A force		Bending force , drawing/weaving force , wave force . (See Budge, p. cvii)

In his example, Budge tells us that the Egyptian word depicted by these glyphs is *remen,* which means "to bear, to carry, to support, to hold up." When we look to the actual word entry in Budge's dictionary for *remen,* we see that based on our established criteria, it qualifies as an enumerating definition for the bent arm glyph. In this case, it is not the enumerating glyphs that are being defined, but rather the bent arm glyph, so the enumerating glyphs appear first in the word. As we suggested earlier, Budge tells us that the number three as a trailing determinative indicates the concept of multiplicity or plurality. In this case, the number confirms for us that the enumerating definition is meant to include all three leading glyphs.

ENUMERATING DEFINITION OF THE BENT ARM GLYPH (REMEN)

A force	𓂋 𓂢 𓈖 𓏪	Bending force 𓂋, drawing/weaving force 𓏠, wave force 𓈖, followed by the bent arm glyph 𓂢 and the number 3 determinative 𓏺 𓏺 𓏺. (See Budge, p. 425a)

What is of particular interest about this example is the approach that Budge takes to the definition of the bent arm glyph and the word *remen.* Budge specifically tells us in his List of Hieroglyphic Characters that the glyphs he presents for the bent arm glyph are taken from the word *remen.* In this instance, Budge shows that he clearly understands that the first three glyphs of the word define the fourth, and he explicitly says so with his bent arm glyph entry. So in his own way, Budge has directly affirmed our concept of an Egyptian defining word. By chance, he has also affirmed our concept of an enumerating definition with the same word entry; the word *remen* ideographically defines the concept of a force by using examples of three specific types of forces.

We can see from all of the preceding examples that our presumption is a correct one; the Egyptian language *does* seem to reflect correct knowledge of the key concepts of astrophysics. These definitions are direct, simple, and consistent, and they are expressed in terms of glyphs whose ideographic meanings are easily confirmed. Likewise, they conform in several key respects to descriptions, words, and drawings from Dogon cosmology. They convey the same quality of information that we find in corresponding definitions from Einstein and express it in the same fundamental terms. These examples support a view of Egyptian hieroglyphs as a consciously and carefully composed language, one that seems to have been intended to convey serious information about the component structures of the universe and of matter.

MEANINGS OF EGYPTIAN GLYPHS
Interpreted from Defining Words

Glyph	Meaning	Terms and Budge Page Reference
	Light	*aakhu,* p. 23a
	Source, limit, result	*khet,* p. 525a
	Space	*pet,* p. 255b
	Bending/warping of gravity	*ref,* p. 423b
	The path of an orbiting body	*shenu,* p. 743b
	A unit of time (year, month, season)	*renp-t,* p. 427b; *abt,* p. 40b; and *ses,* p. 696ab
	An orbit, orbit of the earth around the sun	interpreted in prior discussion, see p. 68
	The bending force of gravity	*Ra,* p. 417b; *aether,* p.101b
	Time	p. cxxi; *atru,* p. 100a
	Time	p. cxlv; *sab,* p. 588b
	Increase in mass caused by acceleration	*s-khakh,* p. 689a
	The strength of or a piece of time	p. cvii, *s-skhakh* p.689a
	Place, to grow	*b,* p. 197a; and *rutu,* p. 421b
	The electromagnetic force	*aun,* p. 34a
	An electron orbit	p. cxxiii; and *aun,* p. 100
	Effect of gravity	*sher,* p. 749b

Glyph	Meaning	Terms and Budge Page Reference
	Weak nuclear force	*sen,* p. 603a
	Strong nuclear force	*sennu,* p. 604b

From *An Egyptian Hieroglyphic Dictionary.*

SEVEN

EGYPTIAN GLYPHS, WORDS, AND DEITIES

Based on examples presented in the previous chapters, we now are familiar with the ideographic form of a fairly wide range of Egyptian words relating to the structure of matter. In our discussions of these examples, we have seen a variety of recurring features of the Egyptian language that we can use to help identify and understand other Egyptian words. Now it is time for us to step back, take a broader look at these examples, and try to place them, as a group, into a coherent framework. However, before we can do that, we need to try to make sense of the overall pattern of language we see before us. In essence, we need to understand how individual glyphs and words work in the Egyptian hieroglyphic language.

As we embark on this next phase of analysis, it is important to remember that our initial approach to the Egyptian language was not a blind one. We began with a well-defined set of Dogon words and symbols, whose symbolic meanings were simply overlaid on matching Egyptian words and glyphs. The legitimacy of this process is borne out by the many persistent resemblances that color most other aspects of Dogon and Egyptian culture. Support for the soundness of this approach is also found in Sir E. A. Wallis Budge's *An Egyptian Hieroglyphic Dictionary,* which provides us with consistent validation of correspondences between Dogon and Egyptian words and symbols. The *Dictionnaire Dogon* by Genevieve Calame-Griaule also specifically validates Budge's definitions and pronunciations of Egyptian words that are similar to those in the Dogon language. So from the very start, we can be confident that our interpretations are based on a cross-confirming set of independent sources.

The more we work with Egyptian glyphs and defining words, the more aware we become of a recurring pattern involving these glyphs and words. Although only a subset of Egyptian glyphs can be said to have a directly assigned phonetic value, examples show that this subset seems to include many of the same glyphs that we believe define key components in the structure of matter. First among these is the flying goose glyph , which we associate with the atom—the counterpart of the mythological po of the Dogon. In Budge's view of the Egyptian language, this glyph carries the pronunciation of "pa,"[1] the same pronunciation as the Egyptian word *pa,* which means "to be, to exist."[2] There is a likely corresponding Egyptian god named Pau, who is defined by Budge as a "god of existence."[3] A similar relationship between a glyph's phonetic value and a deity name appears with the oval glyph , which, based on its appearance in various Egyptian words, can be seen as a symbol for protons, neutrons, and electrons. Budge assigns a phonetic value of "sen" to this glyph, a value that is a likely correlate to the Dogon word *sene.*[4] He defines a similar Egyptian word, *Senu,* which refers to "a company of gods."[5]

The pattern that is seen to repeat in these initial examples begins with a glyph that appears to be symbolic of a component in the structure of matter. The assigned phonetic value of the glyph is similar to the pronunciation of the name of an Egyptian deity (or class of deities) whose traditional role in Egyptian mythology correlates to the proposed component of matter we suggest is symbolized by the glyph. We can test the consistency of this pattern by looking to another Egyptian glyph, in this case, the wave glyph . Budge tells us that this glyph is pronounced "n"[6] (or "nu" in plural), a phonetic value that we know relates to the Egyptian mythological concept of water and to the Dogon prefix *nu-,* which Marcel Griaule and Germaine Dieterlen also specifically relate to water in *The Pale Fox*. There is also an Egyptian word *Nu*, which Budge defines as "the deified primeval waters whence everything came."[7] This recurring pattern points to a possible convention within the Egyptian language, one that might assist us in identifying specific glyph symbolism that might relate to the structure of matter. The relationship between the role of the deities in our examples and the symbolic cosmological meaning of the associated glyphs seems clear: Pau, the god of existence, corresponds to the glyph we associate with the atom, and Senu, the company of gods, corresponds to protons, neutrons, and electrons—a class of particles that we identify as components of the atom. By this view, Nu, the deified concept of the primeval waters, would correspond to massless waves—the underlying primeval source of matter.

Yet another example of this same pattern is demonstrated by the serpent glyph 𓆓, which Budge pronounces "tch."[8] The serpent is the figure we associate with the completed Calabi-Yau space in string theory—the eighth stage that follows seven stages of vibration, which the Dogon associate with the concepts of tearing and the formation of the Divine Word. Budge defines the phonetically similar Egyptian word *Tchet* as "the Divine Word," which he interprets as the concept of "speech deified."[9]

The idea that the various Egyptian gods and goddesses may have symbolized component stages of matter would seem like a logical outgrowth of Egyptian cosmology as we know it. These deities, commonly referred to as *neters,* are defined by their very name as bearing a relationship to the goddess Neith, who is in turn explicitly described as the mother of all the Egyptian gods and goddesses. Neith's primary role in Egyptian mythology is overtly defined as the weaver of matter. It would only make sense that deities who bear a direct relationship to the weaver of matter would themselves symbolize aspects of the process by which matter is formed.

There is some suggestion that a third set of relationships may exist within the Egyptian hieroglyphic language based on this same group of phonetic values—the direct phonetic values of the Egyptian glyphs. These are to a class of words that I call mythic pronouns. These include personal pronouns such as *I, you, he, she, we,* and *they* but may also include other pronouns such as *someone, anyone,* and *no one.* Because the parallels to these words are often less than exact and the inclusion of yet another level of word correspondences here would serve mainly to complicate the argument, I have chosen to omit this aspect of the Egyptian language from the discussion. However, it is possible that the phonetic values of these matter-related glyphs may have also been tagged to this class of pronouns, which are words representing a part of speech that is unlikely to disappear from a language, even over long periods of time.

In light of the apparent convention of the Egyptian language that seems to associate the pronunciation of a glyph with the name of an Egyptian deity, it seems appropriate to ask the question: If this pattern of correspondences holds true, then should we not be able to simply read through Budge's List of Hieroglyphic Characters, select the glyphs with directly assigned phonetic values, find the god or goddess whose names are pronounced like the glyphs, correlate the deity's role to a process in the formation of matter, and then simply lay out a list of progressive components of matter? Would that not be the most direct way to identify which glyph represents which element within this symbolic structure, and wouldn't the traditionally defined role of the associated god or goddess serve to confirm the scientific meaning of the glyph? If we pause for a moment to pursue this line of thought, we find that it takes only a brief period of straightforward research with Budge's dictionary to answer these questions with an emphatic yes! The glyph phonetic values *do* seem to lead us consistently to the names of Egyptian deities and to related component processes of matter. If we carry this line of research just a bit further, the immediate result is the table on pages 88, 89, and 90, which is meant to correlate Egyptian glyphs based on their phonetic values with the names of known

Egyptian deities or deified concepts, each as a component stage of matter as defined in Dogon cosmology and by modern science. The columns across the table include the phonetic value of a glyph or set of similarly pronounced glyphs, the similarly pronounced name of an Egyptian deity or deified concept, Budge's definition of that deity or deified concept, the likely stage of matter that is represented, and the glyph or glyphs that carry the phonetic value. The rows of the table from top to bottom represent the various stages of matter, whose sequence is dictated by the stages of matter as they are known both in science and mythology. The glyphs are correlated to stages of matter based on the traditional role of their associated deity or deified concept—often the same component stages that we have already surmised for the glyphs based on other evidence—and are organized according to the known sequence of the stages of matter. The table on pages 88, 89, and 90 titled Correspondence of Egyptian Glyphs—Structure of Matter Overview, will be described in further references as the Structure of Matter Overview.

The general sequence of the table is governed by the hierarchy of mythological components of matter in Dogon cosmology as described in *The Science of the Dogon*. This sequence is demonstrably the same sequence of components known to modern science. However, additional logical constraints are imposed upon the table by symbolic relationships known to exist in Egyptian cosmology. For example, we know that the processes of creation begin in Egypt with the concept of nonexistence coming into existence, as symbolized by the dung beetle *kheper*—an insect that the Dogon consider to be a creature of the water. These same processes can be said to end with *Pau,* the Egyptian god of existence who corresponds to the mythical Dogon atom, called the *po*. The specific sequence of glyphs that fall between these two points is driven primarily by inferences made based on a variety of Dogon, Egyptian, and scientific references.

The major divisions of the table reflect three-way agreement between Dogon and Egyptian cosmology and modern astrophysics. Creation begins with massless waves that are believed to exist far below our own plane of being. It proceeds to the Dogon second world or Egyptian Underworld, conceptualized as a Dogon egg that corresponds to the Egyptian Tuat and string theory's Calabi-Yau space. For the Dogon, existence, in the form of particles and atoms, is expressed in the context of a third world that is characterized by cosmological keywords such as *sene* and *po,* which correlate to the Egyptian words *sen, Pau,* and *pau-t.*

Some subsections of the table are organized according to specific Dogon descriptions. For example, the Dogon nummo fish drawing, which we will discuss in detail in the next chapter, defines a series of cosmological shapes in specific relationship to each other—shapes that closely resemble Egyptian glyphs. These well-defined relationships dictate a proscribed sequence for the corresponding Egyptian glyphs. Likewise, events surrounding the formation of the Calabi-Yau space are also dictated by descriptions in Dogon cosmology that are in close agreement with those of string or torsion theory. For example, we know that vibrations characterized by the Dogon as rays of a star inside an egg precede the completion of the Calabi-Yau space, which is conceptualized as a spiraling coil. Such definitions provide a kind of conceptual framework upon which the Egyptian glyphs have been carefully placed.

Although the table of phonetic values, deities, and glyphs presented above constitutes only a first-level overview of what ultimately may prove to be a much more expansive symbolic plan, we can see from its structure that there is a coordinated logic to the assignment of symbols and meanings. It seems unlikely that it could be by mere coincidence alone that the phonetic values of glyphs we associate with components of matter agree consistently with the names of appropriate Egyptian gods and goddesses or that the scientific meanings we assign to those glyphs correlate to the traditional mythological roles of those deities. Rather, what the table suggests is that these relationships reflect a consciously designed plan, a suggestion that is firmly supported by the very nature of the various defining word conventions illustrated in the previous examples.

CORRESPONDENCE OF EGYPTIAN GLYPHS

Structure of Matter Overview

Phonetic Value	Deity or Related Word	Budge's Definition and Page Reference*	Component Stage of Matter or Ideographic Reading	Glyphs Carrying the Phonetic Value
khep/ kheper[1]	Kheprer	Creator of the world (p. 543a)	The creative force	
n/nu	Nu[2]	Goddess of primeval waters (pp. 349b, 352b)	Waves/Water	
ma/maa[3]	Maa	The divine seer (pp. 267a, 279b)	Perception	
a/an	An-t	A mythological fish (pp. 56a, 58b, 124a)	Perception, growth	
Am	Amaa	A serpent god (pp. 120a, 121a)	To know, understand	
kh/khet	Khet	God of things that exist (p. 526a)	"Waves raise up"	
b/bu	Bu	A fiend in the Tuat (p. 197a)	Place	
u/ua/uau	Ua	Birth goddess (pp. 144a, 145a)	Growth comes to be	
aha	Aha	A serpent god/ Latus/Siluris Fish (pp. cxxx, 8b, 134a)	Waves "stand up"	
qet	Qet	A mythological serpent (p. 779b)	"Pedestal gives mass"	
t/ta/taa/ taat	Ta	Primeval Earth god (pp. 815a, 816a)	Mass exists/Fire	

*From *An Egyptian Hieroglyphic Dictionary.*

1. The word *khep* is written with a single glyph, (see Budge p. 541a).
2. *Nu* is the plural of *n,* which is the phonetic value of the wave glyph, (see Budge, p. 349a).
3. The word *maa* is written with a single glyph, (see Budge, p. 266b).

Phonetic Value	Deity or Related Word	Budge's Definition and Page Reference	Component Stage of Matter or Ideographic Reading	Glyphs Carrying the Phonetic Value
t/ten/tennu	Tena	God of three-quarters (pp. 837a, 881b)	Waves separated in two	
sa/sba/saneb	Sab	Wolf-god or Jackal god (pp. 583ab, 588a)	Calabi-Yau space forms	
s/sa	Saa	God of knowledge (p. 583a, 588a)	Bending/binding begins	
tu/tua/tutu	Tua	God of circumcision (p. 823b)	Vibration/Wind	
	Tua-t[4]	The underworld (p. 871b)	Vibrations in Calabi-Yau space	
u/ua/uau	Ua	The One God (p. 153a)	Growth becomes	
u/ur	Ur	A Great God (pp. xcviii, 174a)	Coiled thread grows	
Kha	Kha	The god of "increased mass?" (p. 526a)	Increased mass	
u/un	Un	God of existence (pp. 164a, 165a)	Coiled thread exists	
qa/qua	Qau	God of creation (p. 761b)	Mass becomes raised	
t/tem	Tem	The creator of heaven and Earth (p. 834a)	Mass complete/Earth	
	tem	To complete (pp. 815a, 833b, 834a)		
ar/ari	Arit	A division of the Tuat (p. 130b)	Definition of completed mass	
ab	Ab	(Heart?)[5] god (p. 37b)	Calabi-Yau space complete	

4. Budge defines to mean "Tuat" (see Budge, p. cxxv).

5. The word *ab*, which in another form can mean "heart," is written with a single glyph, (see Budge, p. 37b).

CORRESPONDENCE OF EGYPTIAN GLYPHS (cont.)

Structure of Matter Overview

Phonetic Value	Deity or Related Word	Budge's Definition and Page Reference	Component Stage of Matter or Ideographic Reading	Glyphs Carrying the Phonetic Value
s/set/sett	Set	God of evil (p. 706b)	Bending/binding of mass	
a, aa	aa-t	Two great goddesses (pp. 15a, 107a)	The beginning of existence	
tch/tches/ tchet	Tchet	The Divine Word/ speech deified (p. 913b)	8th stage of Calabi-Yau space	
net	Net	Goddess who weaves matter (p. 399b)	Complex string intersection	
	ntt-t[6]	"To weave" (pp. 299b, 399b)	Simple string intersections	
men[7]/menn-t	Ment	Goddess of woven matter (pp. 297a, 306b)	Woven matter	
ka	Ka/Kaa	God of letters/God of offerings (pp. 782a, 784b, 791b)	Quark	
sen	Senu	A company of gods (p. 604b)	Protons/neutrons	
	senu	Pot, vase, vessel (p. 605b)		
aun/aunn[8]	Aunith	A star goddess (p. 34a)	Electron	
	aunnu	"Nest" (p. 34b)		
p/pa/pau	Pau	God of existence (p. 229ab)	Space and the atom	

6. The glyphs and are not pronounced "net" but are closely associated with the goddess Net.

7. The word *men* is written with a single glyph, (see Budge, p. 296b).

8. Budge assigns a phonetic value of "un," not "aun," to the glyph (see Budge, p. cxxiii).

When we step back and examine our Structure of Matter Overview, certain features stand out that may not have been obvious based on individual case examples. For instance, we can see the four classic mythological elements of matter—water, fire, wind, and earth—laid out in neat succession down the fourth column of the table. Within the context of the table, these keywords are well positioned to reflect four progressive stages in the formation of matter. The table also demonstrates a consistent relationship between the phonetic values of selected glyphs and the names of Egyptian gods and goddesses. These correspondences are important ones, because the defined roles of the associated gods and goddesses appear to emphasize key attributes of the components of matter symbolized by the glyphs and also tend to affirm the symbolic meanings we have already tentatively assigned to many of these glyphs. For instance, the drawing board/loom glyph , which we associate with the weaving of matter, corresponds to the goddess of woven matter; likewise, the sickle glyph , which we relate to the concept of perception, corresponds to the Divine Seer, whose role would seem to express the very concept of perception deified.

According to our overview, the formation of matter proceeds by way of the following events: Unformed waves of matter, which derive from some unknown source of creation, are examined or perceived. Perception causes them to grow and transform, which leads to the growth of mass and vibration, the formation of the Calabi-Yau space, the completion of mass, and the appearance of the coiled thread, as defined in Dogon cosmology. The three types of string intersections cause mass to be bent, bound, and woven—processes that are ultimately responsible for the formation of quarks. These quarks then combine to form larger bodies of mass such as protons, neutrons, and electrons, which in turn form atoms.

The table shows us that the classical mythological element of water corresponds to unformed waves of mass, wind to vibration, and earth to mass. Later discussion will show us that the word *ma,* meaning "to burn up" and represented by the fire glyph , corresponds to the word *maa* and the concept of perception. So we can see that the basic structure of the Egyptian language, even in the abbreviated form of the above overview, conforms in nearly every respect to what we have already learned from string theory, quantum theory, and atomic theory.

There are also consistencies of agreement between the Structure of Matter Overview and what we have already learned about Dogon mythology in *The Science of the Dogon.* The Dogon words *nummo, sene,* and *po* are in close agreement with their correlated Egyptian counterparts *nu maa, sen,* and *pau.* The Dogon mythological concepts of water and clay pots correspond well to their Egyptian mythological and glyph counterparts. Another mythological concept, that of the nest, which the Dogon use to describe the orbit of an electron around the nucleus of an atom, leads us to the Egyptian glyph that is drawn to look both like the Dogon drawing of the nest and the diagram of a typical electron orbit ✢. Our table shows that the Egyptian star glyph ☆ symbolizes the vibrations of a string, in much the same way as the Dogon drawings depict these same vibrations as the successive rays of a star (see Scranton, p. 80). Likewise, the Egyptian hemisphere glyph ⌓ carries much of the same mythological symbolism as the Dogon granary. So we can see that there is a kind of cross-cultural agreement between the Dogon and Egyptian traditions about the shapes of specific characters, their defined symbolic meanings, and the pronunciation of related mythological keywords.

The relationships set forth in the Structure of Matter Overview suggest that Budge's observations about the early assignment of phonetic values to Egyptian glyphs appear to be substantially correct. The consistent correlation between a glyph, its phonetic value, and its placement in the Structure of Matter Overview supports the notion that the phonetic value was assigned as an early feature of written language. However, what Budge did not understand—and in fairness, could have had no way of understanding—was that if the system of language was organized deliberately, then the phonetic values might well have been assigned to specific Egyptian glyphs *at the time of the written language's conception* based on what would have been the broader plan of a consciously designed language. Our table suggests that these values were deliberately imposed as a way of associating specific glyphs with corresponding Egyptian deities, and thereby with corresponding components of matter. This symbolic aspect of the language would then help to explain why we don't find directly assigned phonetic values associated with *every* Egyptian glyph; specific phonetic values would have been meant to serve as a conceptual key to be used to reconcile a larger structure of science with the primordial symbols of language that symbolize it.

Just as the form of the Structure of Matter Overview potentially opens many new doors for discussion about the structure of the hieroglyphic language, it also provides us with important clues into the mindset of the authority that presumably composed the language. The first and most obvious point to be made is that, as we can see from the overall structure of the table, no real effort was made to hide, disguise, or obfuscate any of the scientific details. In fact, from the perspective of an informed observer, the language all but hits us over the head with those details. Symbols that *precisely* match their scientific counterparts are laid out in the appropriate sequence for all who are scientifically aware to see. Prior to the formulation of string theory, researchers were effectively blocked from deducing the underlying meanings of these symbols by their own incomplete understanding of the processes by which matter seems to be formed. However, since the advent of string theory, the only true obstacle to recognition may have been the unfamiliarity of students of the Egyptian language with key diagrammatic shapes of atomic theory and string theory, along with an understandable reluctance to see past long-standing preconceptions about how ancient cultures may have, in fact, emerged.

Among the Egyptian glyph shapes that are strongly suggestive of concepts from modern science, the list begins with the wave glyph, a symbol that commonly appears among many ancient religions and which, of itself, might have aroused suspicions about the underlying scientific nature of ancient myth soon after the appearance of quantum theory. The spiraling coil shape, as it is conceptualized by the Dogon, does not align precisely with the typically conceived looped string of string theory, but it is in close agreement with the suggested windings through which a string is thought to pass on its way to attaining mass, and it may be completely understandable in terms of torsion theory, which postulates the existence of a microscopic vortex at this stage of matter. Three Egyptian symbols that are associated with the goddess Neith as she weaves matter—the three distinct types of string intersections—constitute true identities with scientific diagrams of string theory and so should be immediately recognizable to any observant string theorist. The appearance of these three shapes within the specific context of woven matter is far more than suggestive; it should be a direct cause for investigation. Likewise, the simple mythological statement that matter is woven from threads should surely be the object of immediate curiosity on the part of an observant string theorist.

The Structure of Matter Overview also provides some insight into what may be the true role of gods and goddesses in ancient Egyptian mythology, at least from a symbolic standpoint. These deities seem to be, at this level of interpretation, the keepers of the concepts of science, whose primary symbolic role may be to define, affirm, and preserve the scientific meanings of the glyphs. The Egyptian goddess Nu, who Budge defines as "the deified primeval water whence everything came," could hardly be thought to represent anything but a counterpart to primordial waves of matter. Likewise, Maa, the Divine Seer, must represent a concept similar to perception. Also, Egyptian mythology could not have been more direct or specific about the role of the goddess Neith (Net), the great Mother Goddess who is defined as the weaver of matter. Her role within the hierarchy of the structure of matter is strictly defined in myth and is explicitly confirmed by the glyphs associated with her name and function—the three defined types of string intersection. These kinds of mythological assignments reflect the mind-set of a designer of written language whose intentions may have been to convey in a most open and direct way a structure that is intimately related to science. And so it is only appropriate that the Egyptian god of letters should correlate to quarks in quantum theory and to the 266 fundamental particles of the Dogon mythological structure of matter.[10] Quarks are, of course, the metaphoric counterparts to letters within the language of the structure of matter.

From a design perspective, the choice to associate concepts of science with the gods and goddesses of Egyptian culture is one that speaks to an anticipated need for longevity of tradition. Just as it is not in the nature of a phonetic value to fall completely out of use in a language over long periods of time without surviving evidence, likewise the memory of a deity is not one that fades quickly from the collective mind of a culture. It is evident from these kinds of choices that the theoretical designer of the Egyptian language understood this. Because these choices seem to have been so carefully made, we find ourselves standing before a language that is more than five thousand years old, discussing ancient deities from one of the seemingly most obscure and enigmatic of cultures, and yet still able to reconstruct the encoded details of a complex and coherent scientific system. May I be the first to compliment the architect of this structure and say that someone clearly did a very good job.

Perhaps the last significant detail we should notice regarding this overview of the structure of matter pertains to the organization of key Egyptian glyphs themselves—those that support the backbone of the structure of matter. Our table shows that these are, for the most part, simple, straightforward pictographic ideograms—clear drawings of recognizable objects, symbolizing understandable concepts. We have also seen that the meanings set forth by the Egyptian defining words relating to these glyphs are as simple and direct as the glyphs themselves, even when their meanings involve apparently complex astronomical concepts such as light, time, and gravity. Clearly, it was not the intention of the designer of the language to complicate an issue, even an inherently complex scientific one, without good reason. This is an essential point to be noted, because we can infer from it that many of the more complex Egyptian glyphs and words hold no place in this original structure; they simply do not conform to the mind-set of the authority who theoretically composed it.

EIGHT

THE NUMMO FISH

Now that we understand the various defining conventions that seem to be at work within the Egyptian hieroglyphic language, we are at last in a position to examine what is perhaps the central symbol of the Dogon religion, the nummo fish. In Dogon mythology, the nummo fish stands literally at the doorway to the creation efforts of the god Amma at the point where matter unfolds. Dogon descriptions of the nummo fish are, on their very face, enigmatic in nature[1] and are expressed in terms of mythological keywords that are, in truth, difficult to interpret without recourse to some corroborating reference. Now, however, we have new tools of the Egyptian language at our disposal to help us interpret the Dogon cosmological statements, and among these is our new Structure of Matter Overview.

We have already talked about the major organizational elements that comprise the Structure of Matter Overview. However, if we take an even closer look at these elements, we notice that there is yet another progression of symbols presented in the table—one that involves the images of animals, drawn from the major classes of nature. If we start at the beginning of the table, we see that the original creative force of matter, Khepr er, is portrayed as a scarab or dung beetle —a variety of insect. As we follow the chart to the point of perception, we see that one of the corresponding glyphs of the table is a fish . If we continue to the point of appearance of the Tuat or Calabi-Yau space, the defining glyph takes the form of a jackal, dog, wolf, or fox  —a four-legged animal. By the time we reach the level of the atom, we see that the associated glyph is a goose  .

It is not surprising that these same four categories of symbols are also assigned significance within the Dogon system of cosmology. In the *Dictionnaire Dogon,* Genevieve Calame-Griaule tells us that the Dogon word *ke,* which has a similar phonetic root as Khepre, refers specifically to the dung beetle and in general to all water beetles, and we know that for the Dogon water is the source of creation. Likewise, we are told in *The Pale Fox* that the Dogon symbol of the nummo fish plays a pivotal role in the mythology at the point when creation emerges. The Dogon symbols of the jackal and the fox are assigned virtually the same position and meaning as their Egyptian canine counterparts; the jackal is a symbol of disorder, and the fox is assigned the role of judge. Likewise, in the final stages of the Dogon mythological structure of matter, birds make frequent symbolic appearances, as we have seen they also do in the Egyptian structure. Given the other four-level categorization schemes that are evident within these mythologies (water, fire, wind, and earth; and bummo, yala, tonu, and toymu) our presumption should be that this apparent hierarchy of animal classes exists within the structure of the language for some reasoned purpose.

In this case, however, because of the existence of our table, we can do better than simply surmising some purpose for these mythological symbols; the language itself places the symbols at specific points within the context of the larger structure. We can see that the fish glyph corresponds to the point at which perception causes waves of unformed matter to begin to grow. We also can realize, based on the many preceding glyph examples, that we are now in a position to use Egyptian hieroglyphs to help clarify the meanings of some of the more enigmatic Dogon words. And so if we consider the Dogon word *nummo* in terms of its relationship to the Egyptian language, we find that it is no longer an unknowable concept. We have established that both the Egyptian and Dogon words *nu* refer to primordial waves of water—waves that are explicitly defined in both the Dogon and Egyptian systems as the mythological source of being. Likewise, we know that the Egyptian word *maa* means "to examine or perceive." So the combined words *nu maa,* the likely Egyptian counterpart to the Dogon word *nummo,* can be reasonably understood to refer to primordial waves examined or perceived; in other words, the initiating stage in the scientific process of the formation of matter. The clear implication, which is confirmed by Dogon and Egyptian myths, symbols, and language, is supported by our Structure of Matter Overview, and is reaffirmed by what we know of astrophysics itself, that the Dogon nummo fish drawing will tell us something about unformed waves of mass and how perception causes them to grow.

Based on these symbolic hints, it might make sense for us to pause and review the more obvious elements of the nummo fish drawing, even before we try to reconcile this drawing with Dogon mythological descriptions or any possible parallels from science. We might begin by simply asking ourselves a question: if this figure were truly meant to portray the perception of primordial waves of matter, how would the process work?

Clearly, any interpretation we might choose to overlay on the drawing should be consistent with the hieroglyphic definitions of light, time, gravity, and acceleration set forth in previous examples. Likewise, given the many other consistent parallels between myth and science, we would expect this interpretation to be in close agreement with modern scientific definitions of these same terms. There are several key points that we should keep in mind as we proceed with this discussion. First, both the Egyptians and Einstein's theory of relativity tell us that mass bends time. This means, in effect, that a more massive body can be said to exist in a slower time frame than a less massive body. If time slows down as mass increases, then the converse must also be true; time must speed up as mass decreases. Therefore, a virtually massless particle, such

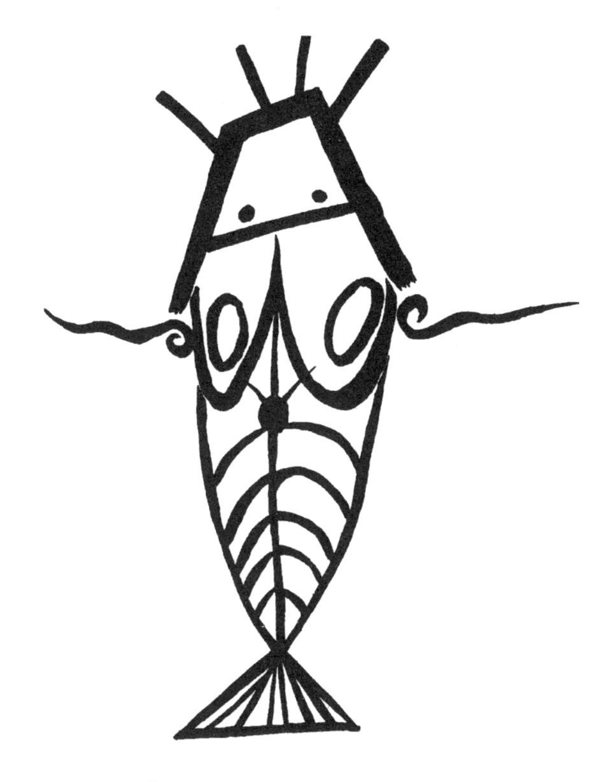

The Dogon nummo fish drawing
(from Griaule and Dieterlen, The Pale Fox, *p. 185)*

as one that might be part of a primordial wave, must exist within an ultrafast time frame when compared to our own experience—perhaps too fast a time frame for us to even measure with any accuracy. If so, then what appears to us through experiment as the random behavior of massless particles might well simply reflect the regular—but for us immeasurable—motions of those particles moving within the context of an ultrafast time frame. From this perspective, we begin to see a possible way to reconcile Einstein's instinctive insistence on a nonrandom universe with the seemingly random results that are consistently borne out by modern-day quantum experiments.

Heisenberg's Uncertainty Principle tells us that when dealing with particles of very small mass, any attempt to observe a particle will necessarily disturb it. This same principle must also hold true for virtually massless particles, whose lack of mass can be alternately expressed as a virtual absence of acceleration. If it is inevitable that a virtually massless particle must be disturbed when it is perceived, then it is equally inevitable that the same particle must be accelerated by that act of perception. Again, Einstein tells us that when we accelerate a body, we slow its time frame, and by slowing its time frame, we give it *mass*. So based purely on Einstein, Heisenberg, and our hieroglyphic definitions of light, time, and acceleration, we find that we are able to explain how a simple act of perception might cause a virtually massless particle to gain mass: perception causes acceleration that translates into a gain in mass. The consequence of this gain in mass is a requisite slowing of time frame, an event that may well move the massless particle out of its nearly infinite time frame and into the effective range of our own perception. And so what previously appeared to us as a primordial wave suddenly transforms itself, as if by some ancient magic, into a tangible particle.

With those points in mind, we now are ready to consider the nummo fish drawing and how it might represent processes related to the perception of waves. Beginning at the tip of the tail of the fish, we could postulate that the bottommost triangular shape represents some underlying source of waves that is unknown and undefined. Within the tail of the fish, we see a series of wavelike lines separated by a predefined midline—perhaps a demarcation of the future duality of matter. The spiked ball that appears in the middle of the body could represent the point of perception or, as Heisenberg would insist, the point of disruption of the wave. Such a disruption would have the effect of accelerating the wave at the point of perception and of slowing its time frame at that point. Because a point in the middle of the wave has been accelerated, it draws up, causing the wave above it to divide and encircle, forming twin pools. These pools might be represented by the Dogon perfect twin pair (the nummo), described by the Egyptian phrase *nu maa* (waves perceived), perhaps correlating to two Calabi-Yau spaces in string theory. After being enclosed, the wave that formerly existed at an ultrafast time frame begins to vibrate, the effect of which is to further slow its time frame and increase its mass. Two serpentlike membranes of matter would be among the by-products of the process—one associated with each Calabi-Yau space. Moving toward the head of the fish drawing, above all these other structures, we see a squared hemisphere similar to a flat-topped pyramid or a Dogon granary, the same shape that I associate in *The Science of the Dogon* with the concept of mass. Beyond the head of the fish are four separate lines that might represent the four quantum forces: gravity, electromagnetic force, weak nuclear force, and the strong nuclear force. These quantum forces are further by-products of this process.

Having considered briefly how the Dogon nummo fish drawing *could work,* we are now ready to review specific Dogon statements about how the drawing *does work* according to the Dogon priestly tradition. These statements are found beginning on page 184 of *The Pale Fox,* chapter 1, section 6, titled "Amma's Work—Development of the 'Second World.'" Griaule and Dieterlen tell us that this process of creation consists of giving volume to a creative force through the successive stages of the drawing. The process begins with what is described as Amma's egg in a ball *(amma talu gunnu)*—the spiked ball at the center of the nummo fish drawing. We will learn later that the Dogon consider this spiked ball to be Amma. As we have found with previous Dogon symbols and drawings, the figure of this spiked ball also exists as an Egyptian hieroglyph; Budge tentatively identifies it in his List of Hieroglyphic Characters as representing a date, a kind of fruit (see Budge, p. cxxiv). For Budge, the Egyptian word for date is *am* or *amm* and is written with the Egyptian glyphs , which can be interpreted ideographically to mean "that which is observed" and constitutes a defining word for the egg glyph. Budge defines the owl glyph to mean "see or behold";[2] the glyph is defined ideographically as , or "that which is (or has been) seen or beheld."[3] The glyphs of the Egyptian word *ama* translate ideographically as "the act of perception."[4] So our hieroglyphic starting point for this creation process agrees with the

Dogon in three respects: the shape of the spiked ball symbol, the pronunciation and meaning of the mythological term *amm* and the name of Amma, and the mythological and ideographic concepts represented by the ball as an initiating egg. Dogon descriptions tell us that Amma produces his creation by opening himself up, a choice symbolized by the figure above the spiked ball, which is said to depict a double door (*ta dine,* meaning "twin door") locked with the "key of the world." According to Budge, the Egyptian word *tati* represents "two sides of a door."[5] Likewise, the word *tat,*[6] which means "emanation," can be written with the single teardrop glyph  , a figure that re-creates the shape of one of the Dogon twin doors. Budge defines a related word, *taa,* as meaning "divine emanation."[7] "Taa" is a defining word for the teardrop glyph and is written with the glyphs  , or ideographically "mass is begun or evoked (made to come to be)" followed by the teardrop glyph and the god determinative. Other Egyptian words with the pronunciation "ta" carry the following meanings: "a mark of the dual" ( ideographically), "fire exists,"[8] "time" ( ideographically), "fire is spoken,"[9] and "world." The word *tah* means "to dip in water," and *Ta-t* is a reference to the "other world," or the Tua-t.[10] So we can see that the Egyptian concepts and ideographic characters that correspond to the Dogon word *ta* reaffirm the appropriate scientific and mythological theme relating to this phase of the formation of matter: the emanating point at which time is initiated, mass starts to form, and the duality of matter begins, symbolized by the mythological keyword *fire.*

When the Dogon speak of Amma's clavicles, in one sense they refer to the collarbone shape formed by the twin doors 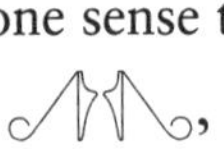 , the peak formed by the perceived wave as it is drawn up by acceleration. The pedestal upon which the wave rises is the Egyptian glyph  , pronounced "qet," which Budge explains as  or ideographically as "earth (mass) given."[11] An entry in Budge's dictionary tells us that the Egyptian word *qet* means "to build, to fashion, to form, to construct."[12]

The Egyptian words that correspond to the Dogon mythological phrases in *The Pale Fox* lead us to words similar to ta, tua, and Tua-t. This is not at all surprising because we already have associated the scientific concept of the Calabi-Yau space with the Egyptian concept of the Tua-t—the same structure that we believe correlates to the circular pools depicted above the twin doors in the Dogon nummo fish drawing. When we look to the Egyptian language for words relating to *tu, tua,* and *Tua-t,* we find the word "Tuat," written, meaning literally "vibrations of mass" and defined by Budge as referring to "a circle in the Tuat."[13] The Egyptian word *tuaut* refers to "hollows, empty places."[14] Dogon descriptions in *The Pale Fox* tell us specifically that Amma's clavicles, now round, became the four extended cardinal points of the earth.[15] Egyptian language echoes these Dogon descriptions when it uses the word *tua-t,* which can also refer to one of the four supports of the sky, and the word *tun,* which means "to extend, to rise up, to lift up."[16]

The Dogon describe the twin doors as being closed on one side (the lower side) to represent the things that Amma will keep to himself; the other side, in which the key is placed (the key being the head of the fish, which is shaped like a flat-topped pyramid) represents what Amma will give up. *The Pale Fox* tells us that the key is like the "germination of the cereal grains," which might explain the similarity of its shape to the Dogon granary (see the figure on page 36). One Egyptian term for granary is *ukher,* which is written, a set of glyphs that we can interpret ideographically as meaning "the coiled thread's source of bending," followed by the chamber glyph. Budge defines the Egyptian word *ukha* as a "whirlwind or storm." It is written and can be understood ideographically as meaning "the growth of acceleration (or increase in mass) becomes vibration." A related Egyptian word, *ukha theb-t,* is the term for the base of a pyramid.[17] And so the growth of matter is associated in several different ways with the growth of grains, and it is therefore symbolized at this stage in the Egyptian hieroglyphic structure of matter both by the granary shape and by a glyph that can be seen to depict the growth of a plant. This unorthodox interpretation derives from the use of the glyph in the words *uab,*[18] which means "plant or flower," and *uah,*[19] which means "to grow." The interpretation is entirely in keeping with the tenor of Dogon cosmology, which describes the formation of matter metaphorically in terms of the growth of seeds into plants. A more orthodox view of the same glyph is that it represents a looped cord, which is similar to the cord interpretation that is traditionally assigned to the spiraling coil glyph .

If the growth of Amma's clavicles came to represent the four cardinal points (the four dimensions of space-time), then we might infer the scientific meaning of "dimension" for the mythological keyword *clavicle.* The concept of the spreading of Amma's clavicles would then refer to the progression of the vibrations of the string as it unwinds within the seven wrapped-up dimensions of the Calabi-Yau space. The Dogon describe this process as Amma moving in a spiral, leaving rays in his wake. This is the very same process that is discussed in greater detail in *The Science of the Dogon* based on Dogon descriptions and drawings from *The Pale Fox.*[20] These descriptions reaffirm the notion of the Calabi-Yau space as a circular space and explicitly define the vibrations within the Calabi-Yau space as individual rays of a star, each of increasing length, whose endpoints can be said to describe a spiral that is later depicted as the coiled thread .

According to descriptions in *The Pale Fox,* this same series of processes associated with the nummo fish and the raising up of matter also have direct bearing on the formation of Amma's *Word.* We can relate the mythological concept of the Word to the serpentlike projections that appear as by-products of the process shown in the nummo fish drawing—structures that may correlate to membranes in string theory. The assignment of the mythological concept of the Word to the symbol of the serpent is explicit within the Egyptian language and is based on the Egyptian word *Tchet,* which Budge defines as "the Divine Word," or "speech deified."[21] It is written with the Egyptian glyphs , which can be read literally as "serpent (or Word) given," followed by the god determinative ( ).

We see by these examples that each of the component elements of the nummo fish drawing is mirrored by an element of the Egyptian language. Dogon keywords lead us to Egyptian words that are pronounced like their Dogon counterparts and are written using glyphs whose shapes correspond to the component shapes of the nummo fish drawing. Also, each Egyptian word carries a set of meanings that is appropriate to a scientific discussion of the formation of matter. The Dogon plainly tell us that the constituent elements of the nummo fish drawing describe a series of events that occurred at the time of creation, outside of space and time, beginning with the fundamental act that initiates the formation of matter—the perception of primordial waves. Each subsequent event follows in the logical sequence we surmised during our preview of the nummo fish drawing, a sequence that could easily describe a transition from waves to particles. Each of the otherwise enigmatic Dogon references from *The Pale Fox* is specifically clarified by the Egyptian language, supported by an Egyptian glyph whose shape replicates the appropriate Dogon symbol, and further defined by sets of Egyptian words whose meanings expand on the related Dogon mythological descriptions. The series of events, as confirmed both by Dogon descriptions from *The Pale Fox* and by the Egyptian language, lead us through the nummo fish drawing, beginning with the perception of primordial waves of matter, to the drawing up of the perceived wave, through the twin doors of Amma's clavicles, through the encircling waves that form the Calabi-Yau space, to the point of the initial vibrations of the string, to completed mass, and on to the ultimate formation of the finished Word, which is depicted both in the Dogon drawing and in the Egyptian language as the symbol of the serpent. Along the way, we discover additional correspondences between Dogon and Egyptian myth and language that confirm the symbolic meaning of the shape of the Dogon granary, define the relationship between the formation of matter and the giving of the grains, and may ultimately assign new mythological parallels—

consonant with those outlined in *The Science of the Dogon*—between the pyramid and this same process.

MEANINGS OF EGYPTIAN GLYPHS
From the Dogon Nummo Fish Drawing

Glyph	Meaning	Reference
	Waves raised up by perception	See the nummo fish drawing
	The growth of mass	See the nummo fish drawing
	Waves separated in two (twin doors)	See the nummo fish drawing
	Membranes in string theory (the Word)	See the nummo fish drawing

See page 99.

NINE

SYMBOLIC STRUCTURE OF THE EGYPTIAN LANGUAGE

The discussion of the nummo fish drawing in the preceding chapter should have opened our eyes to important new insights about the Egyptian structure of matter. With the nummo fish, we selected a single line entry from the Structure of Matter Overview and then expanded its definition by using details from Dogon mythology and the Egyptian language. We found that the Egyptian fish glyph 𓆛 actually defines a process, not a single component of matter, and that this process consists of several smaller component stages, each of which is associated with its own Egyptian glyph. We know that at key stages in the structure of matter, we should expect to find a predictable set of relationships between the phonetic value of the glyph and the name of a related god or goddess. These relationships, by their very nature, invite us to look beyond the initial glyph to other Egyptian words and glyphs that carry a similar pronunciation, just as the Dogon priests explicitly tell us to look for relationships between words based on similarities of pronunciation. The suggestion is that we may learn more about the symbolic concept assigned to the glyph (and the stage of matter it represents) by exploring other phonetically similar Egyptian words.

In the Egyptian language, it is common to find a single concept expressed by two or more different words, each with its own distinct pronunciation and spelling. Previous discussions have shown that we can benefit by examining these alternate Egyptian words and the glyphs they contain. For example, *The Science of the Dogon* first took the approach of comparing these kinds of alternate words while investigating the mythological concepts of weaving and threads. Now, if our goal is to further refine the hieroglyphic definitions of each component stage of matter, these alternate words become a likely source of new information about each symbolic stage.

One point that should be made briefly here—a point that is the subject of larger discussion in *The Science of the Dogon*—is that there may well be a second thread of scientific symbolism running through the Egyptian language, one that follows details of biological reproduction. This thread shares many of the same creation themes, symbols, and keywords as the structure of matter thread, and so our investigations of the Egyptian language may turn up words whose ideographic meanings actually pertain to concepts of genetics, not astrophysics. These words are easily recognized because they often include the twisted rope glyph , a character whose mythological symbolism may be linked to the concept of the double helix of DNA. Although the ideographic meanings of these words do not pertain to our study, their common-usage meanings may well express concepts that pertain to the formation of matter.

It immediately becomes clear that the examination of alternate Egyptian words for a given concept can be productive when we review the Egyptian concept of perception, an act that initiates the formation of matter. One Egyptian word for "perception" is *an,* and, based on our criteria, it qualifies as a defining word for the Egyptian eye glyph (see Budge, p. 123a). The word *an* is written with the bent arm glyph, the wave glyph , the hemisphere glyph , the spiraling coil glyph , and the Egyptian eye glyph . The image of an eye, in this case, the glyph that appears to be defined by the word *an,* would seem like an appropriate choice as a symbol for the act of perception. Budge lists another word *an,* which he describes as "a mythological fish." In fact, based on our criteria, it is an Egyptian defining word for the fish glyph . Its glyphs can be read ideographically to mean "the act that weaves coiled threads," followed by the fish glyph .[1] Likewise, the fish glyph itself carries a phonetic value of "an."[2] A related Egyptian word, *an-t,* means "to know, to perceive." Another Egyptian word *an-t* refers to the Egyptian adze blade, the symbol of the processes by which waves attain mass. When we apply ideographic meanings to the glyphs of the first of these words *an,* they read as follows:

DEFINITION OF THE EYE GLYPH (AN)

Perception		The force or act that weaves mass via the coiled thread , followed by the eye glyph . (See Budge, p. 123a)

Another set of Egyptian words that relate to the act of perception is based on the pronunciation "am." According to Budge, the word *amam* means "to perceive."[3] We can see that a related word, *ama,* carries a similar ideographic meaning. It is written with just two glyphs, the bent arm glyph and the sickle glyph .[4] Based on our emerging system of meanings, these two glyphs together can be interpreted ideographically to mean "the act of perception." On the same dictionary page, Budge defines another word *am,* which is described as a "weaving instrument." He also defines the Egyptian god Amaa as "a serpent god."[5] He lists an even more interesting word, also pronounced "am," that means "to know, to understand,"[6] but it would seem that the ideographic meaning of this word really has to do with the concept of written language. In this case, the wave glyph can be seen to represent the concept of writing (see the Egyptian word *uten,* which means "to write").[7]

Literally, its glyphs read "to place in writing," followed by the hand writing glyph. In the context of this interpretation, the word would use the wave glyph to portray the concept of writing. Based on this definition, the combined ideographic meaning of the word *am* constitutes a virtual statement of purpose for the Egyptian hieroglyphic language.

DEFINITION OF KNOWLEDGE (AM)

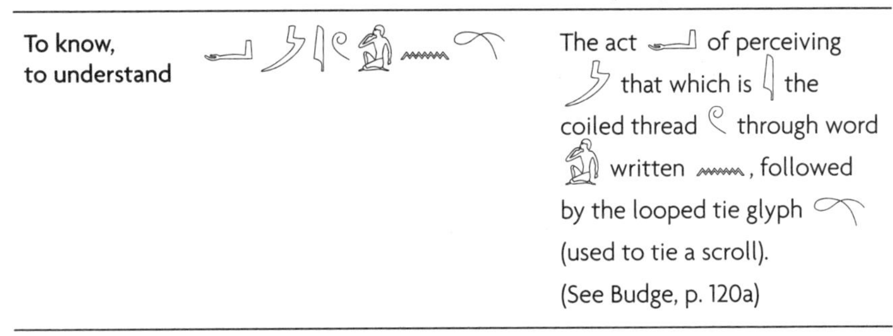

To know, to understand		The act of perceiving that which is the coiled thread through word written, followed by the looped tie glyph (used to tie a scroll). (See Budge, p. 120a)

Our Structure of Matter Overview shows that there is a third set of words relating to the concept of perception that is based on the pronunciation "maa." These words include the name of the Egyptian god Maa, who Budge defines as "the Divine Seer."[8] They also include the word *maa-t,* which means "inspection,"[9] and the word *ma,* which means "to burn up."[10] The second definition is perhaps a reference to the mythological keyword *fire,* the symbol that appears to govern events at this stage of matter. Budge also defines a word *maa,* written with the glyphs, that can be read ideographically as "the perceiving act of waves."[11] In *The Science of the Dogon,* I discuss a related Egyptian concept, *bu maa,* which means "to examine or perceive." It is the likely Egyptian counterpart to the Dogon word *bummo* and the name of the first of four mythological stages of creation.

Based on this first example, the larger pattern we see in the Egyptian language is this: When a glyph relates to a component of matter, its phonetic value leads us to an Egyptian deity whose mythological role defines the concept symbolized by the glyph. The phonetic value also leads us to a set of Egyptian words whose meanings may offer additional insights into the concept symbolized by the glyph. In some cases they correspond to a Dogon mythological keyword, or in other cases their definitions may add subtle nuances of meaning to the defined glyph. The ideographic form of these words may also corroborate the scientific meaning of the original glyph. When we explore alternate Egyptian words, which are pronounced differently but express the same mythological concept, we find that the meanings of these words also may contribute to our understanding of the original glyph.

If we return now to our Structure of Matter Overview, we move past the stage of perception to the next stage, in which threads or strings begin to grow. This is the point at which the perceived wave rises up to form what eventually will be twin Calabi-Yau spaces. This concept is expressed by the Egyptian word *ua,* which is a defining word for the fish glyph. It is written 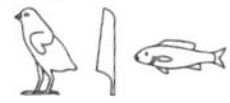 and read ideographically as "growth is" or "growth exists," followed by the fish glyph.[12] In terms of the steps of our Structure of Matter Overview, this constitutes the first stage in the formation of mass, and so it corresponds to Ua, the Egyptian One God.[13] Budge defines another Egyptian word *ua* as meaning "one who becomes eight," which is perhaps a reference to the eight stages of vibration of the coiled thread.[14] There is also an Egyptian goddess named Ua, who is a birth goddess. The glyphs of her name read  , or ideographically, "growth comes to be." A related word, *uatch,* which Budge defines as "a twig, a stick, a column or a support,"[15] is a defining word for the branch glyph  , which is a symbol for wood, that which fire burns, and refers to the drawing up of the wave prior to the formation of the twin Calabi-Yau spaces.

Many of the Egyptian words that carry the pronunciation "ua" are written with the  glyph, a character Budge associates with a pike or a harpoon but that also strongly resembles the shape of a traditional Dogon gate lock. In some ways, the glyph also calls to mind the shape of a fish.

The overall impression it gives is almost one of a fish under construction—or perhaps the conceptual frame upon which a fish is built.

The concept of "waves rising up" leads us to the Egyptian word *kh,* which means "to be high, to rise up," referring to the waters of the Nile, and is written ideographically as "pool of water rises up" .[16] Budge defines the Egyptian god Khet as the god of things that exist.[17] The word *khe-t* also means "fire." It is an Egyptian defining word for the fire glyph and is written , or ideographically, "the source of mass," followed by the fire glyph. Another word *khet* means "grain."[18] The name Qet (which Budge defines as a homonym for *khet*) refers to a mythological serpent, whose ideographic name reads , "pedestal gives mass."[19] The word "pedestal" refers to the vertical portion of the Egyptian teardrop glyph . Another Egyptian word *qet* means "circle"—its glyphs read or "pedestal gives circle," a reference to the circular shape of the Calabi-Yau space that is formed as the wave draws up.[20]

We could continue with this same type of analysis at each component level of matter in our Structure of Matter Overview—starting with the concept expressed by the Egyptian glyph, finding alternate Egyptian words that express the same concept, and then using the meanings of those words to further define the component of matter. We can see that this approach probably will lead us to other glyphs and concepts, each with its own place in the overall structure. These words might offer additional information about processes that occur at a given stage of matter, or they might provide links to other Egyptian gods and goddesses whose mythological roles pertain to the component of matter. Although we already can see that it would be tedious and quite lengthy to present every component stage of matter in an itemized format, the preceding examples show that its net effect would be to expand our Structure of Matter Overview. Each line item would most likely grow to include a set of related words and glyphs. In the end, our modest Structure of Matter Overview would evolve into a much lengthier, more fully defined Structure of Matter table.

We could continue with this same type of lengthy discussion for each component stage of matter in our Structure of Matter Overview—starting with the concept expressed by the Egyptian glyph, finding alternate Egyptian words that are pronounced the same way or express a similar concept, and then using the meanings of those words to further define the component stages of matter. We see that this approach leads us to other glyphs and concepts, each with its own place in the overall structure. These words offer additional information about processes that occur at any given stage of matter and provide us with links to other Egyptian gods and goddesses whose mythological roles may pertain to the component stage of matter. The net effect of this analysis is to expand the Structure of Matter Overview. In effect, each line ultimately produces its own section of a more detailed table.

The organization of this new table, Correspondence of Egyptian Glyphs—Structure of Matter Detail (referred to later as Structure of Matter Detail), is generally the same as for the Structure of Matter Overview, except that the second column now includes additional related words, which may or may not reflect the name of an actual deity or deified concept. Column four of the table for these entries may include my ideographic reading of the Egyptian word or one or more pertinent Egyptian glyphs. If a single glyph appears, it is because I consider the glyph to be defined by the Egyptian word. If multiple glyphs appear, they represent Budge's spelling of the Egyptian word, and I consider that spelling to be pertinent ideographically. For the sake of expedience within the table, these glyphs are given in a single line to be read from left to right, rather than as glyphs written above and below as they would appear in Budge's dictionary. Also, glyphs that appear in parenthesis denote Egyptian characters that are closely associated with the component stage of matter, but not actually pronounced like the phonetic values listed in the first column.

CORRESPONDENCE OF EGYPTIAN GLYPHS

Structure of Matter Detail

Phonetic Value	Deity or Related Word	Budge's Definition and Page Reference*	Component Stage of Matter or Ideographic Reading	Glyphs Carrying the Phonetic Value
		THE CREATIVE SOURCE		
khep/kheper	Khepre[1]	Creator of the world (p. 543a)	The creative force (Dogon "ke")	
	khep	He creates what is (p. 541a)	"source of space"	
	khep-t	A kind of goose (p. 541b)	"source of space and mass"	
		PRIMORDIAL WAVES		
n/nu/nua	Nu	Goddess of primeval waters (pp. 349b, 352b)	Waves/**Water**	
	Nu	Celestial waters; mass of water that existed in primeval times (p. 349b)	"particles and waves" (Dogon **Nu**)	
	Nu[2]	Adze (p. 352a)		
	Nua	A tool of Anubis (p. 352b)	"waves grow"	
	Nunua	To tremble, quake (p. 352b)	"to vibrate"	
	nu-t	Mass of water (p. 349b)		

*From *An Egyptian Hieroglyphic Dictionary.*

1. The word *khep* is written with a single glyph, (Budge, p. 541a).

2. "Nu" is the plural of "n," which is the phonetic value of the wave glyph, (Budge, p. 349a).

Phonetic Value	Deity or Related Word	Budge's Definition and Page Reference	Component Stage of Matter or Ideographic Reading	Glyphs Carrying the Phonetic Value
		PERCEPTION		
m/ma/maa	Maa	The Divine Seer (pp. 267a, 279b)	Perception/Fire	
	ma	To burn up (p. 268b)		
	maa[3]	To examine or perceive (p. 266b)		
	Maa-t	Order (p. 271b)		
	maa-t	Sight, vision (p. 266b)		
	maa-t	Inspection (p. 266b)		
an/ant	An-t	A mythological fish (pp. 56a, 58b, 124a)	Perception, growth	
	an	Cord (p. 58a)		
	an	To turn a glance toward (p. 123a)		
	an	Mythic fish (p. 124a)		
	an-t	The Adze (p. 124a)		
	an-t	To know, to perceive (p. 128a)		
Am	Amaa	A serpent god (pp. 120a, 121a)	To know, understand	
	am	Fire (pp. 6a, 49a)		
	am	Weaving instrument (p. 121b)	Shuttle of a loom	

3. The word *maa* is written with a single glyph, (Budge, p. 266a).

CORRESPONDENCE OF EGYPTIAN GLYPHS *(cont.)*
Structure of Matter Detail

Phonetic Value	Deity or Related Word	Budge's Definition and Page Reference	Component Stage of Matter or Ideographic Reading	Glyphs Carrying the Phonetic Value
	ama	To see (p. 6b)		
	amam	To know, to understand (p. 6a)		
	amam	To perceive (p. 121b)		
	amma	To cause (p. 50b)		
	Amen	The Hidden God (p. 51b)	(Dogon Amma)	
		WAVES RAISE UP		
kh/khet	Khet	God of things that exist (p. 526a)	"Waves raise up"	
	kh	To be high, rise of the Nile (p. 525a)		
	khet	Definition of glyph (p. cxxi)	"The source of mass"	
	khe-t	Fire (p. 526a)	"the source of mass"	
b/bu	Bu	A fiend in the Tuat (p. 197a)	Place	
	bu	Place, house, site (p. 213b)		
	bu	Place, house, site (p. 213b)	"Place of the coiled thread"	
	bu pu	(No definition) (p. 213b)	"Place grows, space grows"	

Phonetic Value	Deity or Related Word	Budge's Definition and Page Reference	Component Stage of Matter or Ideographic Reading	Glyphs Carrying the Phonetic Value
	bu	A sign of negation, not (p. 213b)		
	bu pu ua	No one (p. 213b)		
	bu maa	Place of truth (p. 214a)	"Place grows, perception subsides" (Dogon **bummo**)	
u/ua/uau	Ua	The birth goddess (pp. 144a, 145a)	Growth comes to be	
	ua	The latus fish (p. 152a)		
	ua	To remove, to set aside (p. 152a)		
	uatch	Stick, twig, pillar, support (p. 151a)		
		DELIMITATION POSTS FORM		
aha	Aha	A serpent god Latus/Siluris Fish (pp. cxxx, 8b, 134a)	Waves "stand up"	
	ah	To surround, to enclose (p. 75a)		
	ah	Collar (p. 75b)		
	ahi	Courtyard (p. 74b)		
	aha	Latus fish (p. 132b)		
	aha	To stand (p. 133a)		

CORRESPONDENCE OF EGYPTIAN GLYPHS *(cont.)*
Structure of Matter Detail

Phonetic Value	Deity or Related Word	Budge's Definition and Page Reference	Component Stage of Matter or Ideographic Reading	Glyphs Carrying the Phonetic Value
	aha	Stick, wooden staff, prop (p. 133a)		
	ahau	Delimitation posts (p. 133b)	(Dogon **yala**)	
	ahau	Period of time (p. 133b)		
		WAVES DIVIDE IN TWO AND ENCIRCLE		
qet	Qet	A mythological serpent (p. 779b)	"Pedestal gives mass"	
	qet	To build, fashion (p. 779a)		
	qet	Design, drawing (p. 779b)	The draughtsman's craft	
	qet	Image, likeness, circle (p. 779b)	"Pedestal gives circle"	
	qet	Circle, orbit (p. 780a)		
	Qeti	A god of the abyss (p. 780b)		
	Qetu	The divine potter (p. 779a)		
t/ta/taa/tat	t/ta/taa	Divine emanation (pp. 815a, 816a, 821b)	Mass given	
	ta	A mark of the dual (p. 821a)		
	ta	Staff, support (p. 821a)		
	Ta	God of a circle (p. 816a)		

Phonetic Value	Deity or Related Word	Budge's Definition and Page Reference	Component Stage of Matter or Ideographic Reading	Glyphs Carrying the Phonetic Value
	Ta	Primeval Earth god (p. 816a)	Mass exists	
	ta	Earth, world (p. 815a)		
	ta	Time (p. 815a)	 "Fire is spoken"	
	ta	Fiery (p. 817b)		
	ta-t	Emanation (p. 821b)		
	ta-t	Room, chamber (p. 821b)		
	Ta-t	The other world, Tuat (p. 818a)		
	ta Tuat	Land of the other world (p. 816b)		
t/ten/tennu	Tena	God of three-quarters (pp. 837a, 881b)	Waves separated in two	
	ten	Vase, vessel, bucket (p. 881b)		
	ten	(No definition) (p. 880b)	 "Mass is woven"	
	tenn	To split (p. 881a)		
	ten-t	Chamber (p. 881b)		
	tennu	Border, boundary (p. 881b)	(Dogon **tonu**)	
CALABI-YAU SPACE FORMS				
sa/sba/ saneb	Sab	Wolf-god or Jackal-god (pp. 583ab, 588a)		
	sab[4]	Door, pylon (p. 588b)		

4. The word *sab* is written with a single glyph, (Budge, p. 588a).

CORRESPONDENCE OF EGYPTIAN GLYPHS *(cont.)*
Structure of Matter Detail

Phonetic Value	Deity or Related Word	Budge's Definition and Page Reference	Component Stage of Matter or Ideographic Reading	Glyphs Carrying the Phonetic Value
	Sabu	Wolf/jackal guides of the Tuat (p. 588a)		
s/sa	Saa	God of knowledge (pp. 583a, 588a)		
	sa	Son (egg) of the heart (p. 583b)		
	sa	Fire (p. 589b)		
	saaa	To know		
	saai	To know by sight, to recognize		
VIBRATIONS BEGIN AND MASS INCREASES				
tu/tua	Tua	God of circumcision (pp. 823b, 870a)	Vibration/**Wind**	
	tua	Air, wind, breath (p. 823b)		
	tuan	You, your (p. 824a)		
	Tuat[5]	A circle in the Tuat (p. 872a)		
u/ua/uau	Ua	The One God (p. 153a)	Growth becomes	
	ua	One who becomes eight (p. 153a)		

5. Budge defines to mean "Tuat" (Budge, p. cxxv).

Phonetic Value	Deity or Related Word	Budge's Definition and Page Reference	Component Stage of Matter or Ideographic Reading	Glyphs Carrying the Phonetic Value
u/ur	Ur	A Great God (p. xcviii, 174a)	Coiled thread grows	
	ur	A violent wind (p. 174a)		
	ur	Flame, fire (p. 174a)		
	ur	Lake (p. 174a)		
	urr	To increase, to grow (p. 171a)		
	urit	Chamber, hall, pylon (p. 174b)		
	Urit	A town in the Tuat (p. 174b)		
	Urrt	A name of the Other World (p. 174b)		
Kha	Kha	The god of "increased mass?" (p. 526a)	Increased mass	
	Kha	Chamber (p. 526b)		
	Kha	Substance of the body (p. 528a)		
	Kha	A lake in the Tuat (p. 526b)		
	Khakha	Stars (p. 528a)		
	Khau	Cord (p. 527b)		
		COILED THREAD EXISTS		
u/un	Un	God of existence (pp. 164a, 165a)	Coiled thread	
	un	To be, to exist, to become (p. 164b)	"Coiled thread wavers"	

CORRESPONDENCE OF EGYPTIAN GLYPHS *(cont.)*
Structure of Matter Detail

Phonetic Value	Deity or Related Word	Budge's Definition and Page Reference	Component Stage of Matter or Ideographic Reading	Glyphs Carrying the Phonetic Value
	un-t	Rope, cord (p. 167a)		
	unu-t	Hour, time (p. 167a)		
	unun	To tremble, (p. 167a)	Vibrate	
	unun	To do work in a field (p. 167a)	(Matches Dogon definition)	
qa/qua	Qau	God of Creation (p. 761b)	Mass becomes raised	
	Qa, Qait	The high place where the god of creation stood (p. 761b)		
	Qab	Windings of the Kha (p. 763a)		
	Qa	(No definition) (p. 760a)	"Earth raised"	
	Qa	(No definition) (p. 760b)	"Earth becomes raised"	
	Qa	A title of Temu (p. 761b)		
	Qab	Windings, coils, folds of serpent (p. 763a)		
	qa-t	High land, banks above river (p. 761a)		
	qai-t	Land high above surface of Nile (p. 761b)		

Phonetic Value	Deity or Related Word	Budge's Definition and Page Reference	Component Stage of Matter or Ideographic Reading	Glyphs Carrying the Phonetic Value
		MASS IS COMPLETE		
t/tem	Tem, Temu	Creator of heaven and earth (pp. 815a, 834b, 835a)	Mass complete/**Earth**	
	Tem	An aged god (pp. 834a, 834b)	Full, dual, of the two hands	
	Tem	To make an end of (p. 833b)		
	Temm	To finish, to complete (p. 834a)		
	Tema	To bind together (p. 836a)		
	temt-ta	All (p. 880a)		
	temui	Full, dual (p. 834a)		
	temau	All, complete (p. 834a)	(Dogon **toymu**)	
ar/ari	Arit[6]	A division of the Tuat	Definition of completed mass	
	ar[7]	To go up, to ascend (p. 129a)		
	ar	To complete, to finish (p. 129b)		
	ar	Steps, staircase (p. 129a)		
	ari	A kind of fish (p. 130b)		
	ari	Breeze, wind (p. 130b)		

6. The Arits were seven in number, and each was in charge of a doorkeeper, a watcher, and a herald (Budge, p. 130b).

7. The word *ar* is written with a single glyph, (Budge, p. 129a).

CORRESPONDENCE OF EGYPTIAN GLYPHS *(cont.)*
Structure of Matter Detail

Phonetic Value	Deity or Related Word	Budge's Definition and Page Reference	Component Stage of Matter or Ideographic Reading	Glyphs Carrying the Phonetic Value
	arit	Judgment hall (p. 130b)		
	ar-t	Uraeus (p. 130a)		
CALABI-YAU SPACE IS COMPLETE				
Ab	Ab[8]	(Heart?) god (p. 37b)	Calabi-Yau space complete	
	ab	Draughtsman (p. 38a)		
	Aba	To open (p. 39a)		
	ab-t	(No definition) (p. 37b)	"Heart of matter enclosed"	
	ab-t	A walled enclosure (p. 38a)		
	ab-t	Scepter (p. 38b)		
	ab-ab	Image (p. 37b)		
CALABI-YAU SPACE TEARS/THE WORD IS SPOKEN				
s/set/sett/sti	Set	God of Evil (p. 706b)	Bending of mass	
	set	Encircled (p. 706b)		
	set	To break (p. 706b)		
	set	Thread, string, cord (p. 628a)		
	setch	To break open (p. 716a)		

8. The word *ab* is written with a single glyph, (Budge, p. 37b).

Phonetic Value	Deity or Related Word	Budge's Definition and Page Reference	Component Stage of Matter or Ideographic Reading	Glyphs Carrying the Phonetic Value
a, aa	aa-t	Two great goddesses (pp. 15a, 107a)	"The beginning of existence"	
	a	Roll (scroll/ document) (p. 106b)		
	aa	Door (p. 107ab)		
	aa	Spacious (p. 107b)		
	aa-t	House, abode (p. 107a)		
	aat	Pale (p. 113a)		
	aatch	Paleness (p. 113a)		
tch/tches/ tchet	Tchet	The Divine Word/ speech deified, (p. 913b)	8th stage of Calabi-Yau space	
	tches	To cut, to divide (p. 911b)		
	tchesef	Fire (p. 911b)		
	Tcheses	Goddess who divides (p. 911b)		
	Tchet-s	Serpent goddess, reborn daily (p. 913a)		
	tchet	To speak (p. 913a)		
	tchet	Permanent, abiding, established (p. 913b)		
	tchet	Bodily form (p. 893a)		

CORRESPONDENCE OF EGYPTIAN GLYPHS *(cont.)*

Structure of Matter Detail

Phonetic Value	Deity or Related Word	Budge's Definition and Page Reference	Component Stage of Matter or Ideographic Reading	Glyphs Carrying the Phonetic Value
	tchet-t	Word (p. 913a)		
	tche-t	Place, house, abode (p. 893a)		
	tchet-t	Culmination of a star (p. 913b)		
MATTER IS WOVEN				
net/ntt/ ntt-t	Net[9]	Goddess who weaves matter (p. 399b)	Complex string intersection	
	nt-t	To weave (p. 399b)	Simple string intersections	
	ntt-t	Cord, thread (p. 399b)	,	
	ntt	That which is, everything which is (p. 399a)		
men/menn-t	Ment	Goddess of woven matter (pp. 297a, 306b)	Woven matter	
	men[10]	To be permanent, fixed, stable (p. 296b)		
	mena	A vase or pot (p. 301b)		
	mensh	Cord, tie, bond (p. 305a)		

9. The glyphs do not carry a phonetic value of "net" but are closely associated with the goddess Net.

10. The word *men* is written with a single glyph, (Budge, p. 296b).

Phonetic Value	Deity or Related Word	Budge's Definition and Page Reference	Component Stage of Matter or Ideographic Reading	Glyphs Carrying the Phonetic Value
		QUARKS ARE FORMED		
k/ka	Ka/Kaa	God of letters/ God of offerings (pp. 782a, 784b, 791b)	Quark	
	ka	To bow (p. 791b)	"Collected bodies of the weak force"	
	ka	To be high; Earth raised (p. 783b)		
	ka-t	Collected mass (p. 791b)		
		PROTONS, NEUTRONS, AND ELECTRONS ARE FORMED		
sen	Senu	A company of gods (pp. 603a, 604b)	Protons/neutrons	
	sen	To bow, to pay homage (p. 603a)	(Dogon "bow the head")	
	sen	To open (p. 604a)		
	sen	Clay (p. 604b)	(Dogon "clay")	
	senu	Pot, vase, vessel (p. 605b)	(Dogon "pot")	
	sennu	Image (p. 604b)	(Dogon "image")	
au, aun, un	Aunith	A star goddess (p. 34a)	Electron	
	aunnu	Nest (p. 34b)	(Dogon "nest")	
	aun[11]	To open, to make open (p. 34a)		

11. Budge assigns a phonetic value of "un," not "aun," to the glyph (Budge, p. cxxiii).

CORRESPONDENCE OF EGYPTIAN GLYPHS *(cont.)*
Structure of Matter Detail

Phonetic Value	Deity or Related Word	Budge's Definition and Page Reference	Component Stage of Matter or Ideographic Reading	Glyphs Carrying the Phonetic Value
	aun	To be robbed (p. 115b)	(Dogon "stolen")	
	au-t	Staff, stick with curved end (p. 114b)		
		THE ATOM IS FORMED		
p, pa	p, pa	This (p. 229a)	(Implies "space" in Dogon and Buddhist cosmology)	
	pa	To be, to exist (p. 230b)		
	pa	To fly (p. 230a)		
pau	Pau	God of existence (pp. 229b, 231a)	Atom (Dogon Po)	
	pau-t	The matter or material of which anything is made (p. 231a)		
	Pauti	The primeval god who created himself and all that is (p. 231a)		

The Structure of Matter table tells the full story of the formation of matter, from nonexistence and the first perception of massless waves to the formation of the completed atom. The table is organized in sections, each of which constitutes a major stage in the formation of matter. This organization was accomplished by the use of key glyphs that shepherd us through related stages of the process. For example, we can see that Egyptian defining words for the fire glyph appear at various stages of the table, beginning at the point of perception and continuing to the point where the coiled thread exists. Likewise, we see that the fish glyph makes its appearance at the point of perception and is in evidence through to the point where waves divide and encircle to form the Calabi-Yau space. These references define a component range for the fish glyph that is comparable to that implied by the elements of the nummo fish drawing. The hemisphere glyph, a symbol for mass itself, ushers us through nearly all levels of matter. It makes its first appearance just after the act of perception, at the point where waves raise up to create space and mass, and sustains itself through the formation of protons, neutrons, and electrons.

We can see this same kind of pattern exhibited by the courtyard glyph , which in the context of our table represents pylons or chambers (wrapped-up dimensions) within the Calabi-Yau space. Appropriately, this glyph appears first at the point where delimitation posts form—the formative stage of matter that causes the perceived wave to encircle. References to this glyph continue through the point where the finished Calabi-Yau space is torn by the coiled thread. Glyphs that we take to represent wind, vibration, vibrations of the coiled thread, and increased mass make their appearance at points in the table that seem to describe the vibrations of the coiled thread and the growth of mass within the Calabi-Yau space.

In most cases, if an Egyptian glyph appears in the table at two or more different stages of matter, the suggestion is that these stages belong together conceptually as well as sequentially within the table. So, one of the apparent purposes of the multiple Egyptian words that—by our system—define these glyphs may be to trace the boundaries of specific processes and concepts within the larger structure of matter. In this way, these glyphs and their associated keywords help us organize the mythological components of matter into their proper sequence.

Of the more than fourteen hundred Egyptian glyphs defined by Budge, only a relative handful are represented in the Structure of Matter table, and these are drawn primarily from the subset of Egyptian glyphs that carry direct phonetic values. When we look to the individual words that provide supporting references for the table, we see that they, too, are predominantly written with characters found within this same core group of glyphs, despite wide differences in the pronunciations and meanings of these supporting words. We also notice that the Egyptian words that form the Structure of Matter table tend to be very simple in form—in most cases they are written with two, three, or four glyphs. The table includes none of the more lengthy Egyptian words, which can sometimes be written with upward of a dozen glyphs. In several instances, Budge defines a concept that relates to the structure of matter using only a single ideographic glyph.

The Structure of Matter table defines a variety of keywords that have significance in Dogon cosmology, starting with those for the four creative stages of bummo, yala, tonu, and toymu. These words, which seem to signify distinct stages in the growth of mass, correspond to Egyptian words that are pronounced similarly and carry meanings appropriate to Dogon symbolism. What's more, based on the Dogon model, we find these words laid out in proper mythological sequence, beginning at the point where waves raise up and continuing to the point where mass is complete. We can infer from entries within the table that the Dogon god Amma may actually symbolize the concepts of perception and knowledge. In the Egyptian hieroglyphic language, the word *maa* implies "perception," and the word *am* represents "knowledge." Such definitions are entirely in keeping with Dogon mythological descriptions relating to the god Amma, who is defined as the creative force that underlies matter and biological reproduction and as the agent by which these creative processes are known. The Structure of Matter table also affirms the meaning of the Dogon word *nummo* in similar fashion by pairing it with the Egyptian phrase *nu maa,* which means "waves perceived," and then confirms each subsequent stage of the nummo fish drawing with appropriate Egyptian words and glyphs, whose shapes match requisite Dogon drawings. Each of the key mythological shapes described here and in *The Science of the Dogon* and defined by various Dogon drawings and descriptions  appears within the table in the form of an Egyptian glyph, each makes its appearance in the table at an appropriate stage of matter, and each carries a value that corresponds to the defined meaning of its counterpart in Dogon myth.

Following this pattern, we can look to the Structure of Matter table to explain the meanings of a variety of different symbols from world mythology. The first of these are the classic mythological symbols of water, fire, wind, and earth. Entries in the Structure of Matter table suggest that these keywords correspond to the four major phases in the formation of matter: primordial waves, perception, vibration, and mass. We have already spoken about the term *One God,* and, when we look to the Structure of Matter table, we see that the word *one* refers to the first of the eight stages of the Calabi-Yau space as defined by Dogon mythology—seven stages of vibration and one of tearing. (Our impulse might be to associate such references with the eight paired emergent gods of the Egyptian Ennead. However, we will see in later chapters that these gods, which emerge in pairs rather than units, correlate more precisely with symbols from another episode of Dogon cosmology.)

We can also see based on the table that the concept of divine emanation may refer to the separation of waves caused by the drawing up of a massless wave after the act of perception, which is the formative step that initiates the duality of matter. Likewise, it becomes clear from the position of other entries in the table that the Dogon and Egyptian mythological concepts of an other world or underworld (actually, according to Dogon myth, the second of three worlds) refers to the Calabi-Yau space—the space within which the coiled thread vibrates during the formation of mass. The table also supports a relationship between the symbol of the pyramid and the concept of completed mass. This is the same symbolism that I assign to the granary in *The Science of the Dogon,* which suggests that the granary may be the Dogon counterpart to the

Egyptian mastaba or pyramid. The mythological concept of a Divine Word is also clearly referenced in the Structure of Matter table—it refers to the serpentlike shape that is seen to emerge from the completed Calabi-Yau space. It seems likely that this same concept may be the one symbolized both by the Egyptian Uraeus—the sacred asp that is often represented on Egyptian headdresses—and the serpentlike appendages that appear on either side of the Dogon nummo fish drawing.

Perhaps the most important aspect of the Structure of Matter table is that it documents in a clear and direct way the various methods by which the Egyptian language can be seen to support the concepts, words, and symbols of Dogon cosmology. It brings together in one framework likely Egyptian counterparts to each of the central Dogon keywords and symbols, laid out in the same sequence and symbolizing the same series of events that are described by the Dogon myths. In so doing, it underscores the many consistent and compelling parallels that seem to be in evidence whenever we compare Dogon and Egyptian myths and language.

TEN

THE TUAT

The Tuat is one of the more complex and obscure concepts encountered in Egyptian mythology. The name Tuat calls to mind a diverse set of enigmatic and sometimes contradictory images whose actual symbolism may not be entirely clear, even to modern-day Egyptologists. The Tuat was the name of the mysterious Egyptian underworld—a shadowy place of death and resurrection. On one hand, it was a place of light, associated with the morning star, the dawn, and the hours of the day. On the other hand, it was also a place of darkness, ruled by the jackal, an animal that, for both the Dogon and the Egyptians, symbolized the concept of disorder. The Tuat was also closely associated with the god Sab, a jackal or wolf-god, whose name in the Egyptian language means "judge" and who in the form of Sab-ur played the role of the great judge in Egyptian mythology. Sab-res (as Budge lists it, or perhaps more properly, Sab-Smau) was another name for Anpu, or Anubis—the jackal-god of the underworld.[1]

To make matters even more confusing, there seems to be a persistent degree of doubt among Egyptologists as to which animal might be portrayed by various jackal glyphs. Entries in Sir E. A. Wallis Budge's dictionary often hedge the reference and define a given deity as a "Wolf-god or Jackal-god" (see the definition of the word *Sab*).[2] Other researchers believe that in some cases the correct assignment actually should be to the Egyptian dog. If we rely on Dogon mythology to be our guide in these cases, then in regard to words that are associated with the concept of disorder, we see that the correct symbolic assignment is to the jackal; the Dogon myths are explicit about the symbolism. Dogon mythology is also clear in its definition of the role of judge of truth, only in this case the assignment is to the pale fox—an animal similar to the wolf or the jackal that is native to North Africa and to Egypt and whose image strongly resembles those of various Egyptian jackal glyphs. In Dogon culture, the concept of the pale fox is closely associated with symbols of cosmology—so closely, in fact, that Marcel Griaule and Germaine Dieterlen chose to name their anthropological study of the Dogon religion *The Pale Fox*.

The meanings of many of the Egyptian words relating to the word *Tuat* gravitate around images of stars and circles. Budge defines the phrase *tua neter* ("god of the morning" or "star of the morning") as a reference to the star god.[3] One entry in Budge's dictionary defines the word *Tua-t* as "the morning star," a reference that he assigns to the planet Venus.[4] In another entry, the word *Tuat* is defined as "a circle in the Tuat."[5] So it is not at all surprising to find that the glyph structure of these words also revolves around two recurring images—the Egyptian star glyph ☆ and the star within a circle ⍟. The image of a star within a circle is one that already is familiar to us from Dogon cosmology. Dogon myths describe the seven vibrations of their mythological cosmic thread as rays of a star, and the Dogon diagram these vibrations in a key mythological drawing that takes the form of a star within a circle.[6] Because Dogon mythology is explicit in its assignment of these symbols and the corresponding Egyptian references are—at best—enigmatic, it would make sense to consider the Egyptian glyphs first in the context of their known Dogon symbolism.

We understand based on discussions in *The Science of the Dogon* that the vibrations of the Dogon coiled thread are similar to those of strings in string theory, which are described as vibrating through the seven unseen dimensions that compose the Calabi-Yau space. It is as a result of these vibrations and their journey through the Calabi-Yau space that the coiled thread attains mass. Based on these references, the first question we might ask ourselves is whether Egyptian words for the Tuat, or any of the hieroglyphic defining words for glyphs relating to the Tuat, might involve definitions that pertain to the concept of mass, the growth of mass, or the formation of bodies of mass.

Our experience with the Structure of Matter table suggests that if we want to understand a new Egyptian word or concept, we should begin by examining the root phonetic sound of the word and then look to other similarly pronounced words for additional insights into its possible meanings. In the case of the word *Tuat,* the root phonetic value is reflected in the word *tu* or *tua*. There is an entry in Budge's dictionary for the word *tu,* which carries a defined meaning of "to give."[7] However, when we look to the ideographic concepts represented by the glyphs used to write the word *tu*, we realize that a more exact translation might be "to give growth." Another form of the Egyptian word *tu,* which we will discuss later, has more obvious associations with the formation of matter. Budge interprets it to mean "bandlet," an architectural term that means an encircling band. It is written with glyphs that call to mind the weaving of matter. Among similar entries in Budge's dictionary is another related Egyptian word, *tuait,*[8] which he translates as "dawn" or "early morning" and takes the form of an Egyptian defining word for the star glyph (see below). Ideographically, the word *tuait* means "gives growth to mass," followed by the star glyph. It is interesting to note here that the concept of dawn—the first perceived moment of the new day—is defined in terms of the growth of mass, as if to establish a metaphor between times or hours of the day and stages of the Calabi-Yau space.

DEFINITION OF THE STAR GLYPH (TUAIT)

Dawn, morning		Gives growth to mass, followed by the star glyph. (See Budge, p. 870b)

In this same context, Budge lists a word *Tua-t,*[9] which he describes as a very ancient name for the Tuat. His dictionary entry includes several different spellings for the same word, the simplest of which consists of just the single glyph . Another spelling of the word is written with only two glyphs—the star glyph and the hemisphere glyph . Based on our ideographic definition of the star glyph above, we can see that these characters convey the same sense of meaning as the longer word *tuait;* they depict a star defined in terms of mass. The first spelling listed in Budge's entry for the word *Tua-t* includes four characters and constitutes a defining word for the Egyptian town glyph (see below). This glyph corresponds to a Dogon field drawing that resembles one of the typical vibratory patterns of a string in string theory. When we substitute ideographic concepts for each of the glyphs in this spelling of the word *Tua-t,* they confirm the same meaning that has been suggested by previous spellings:

DEFINITION OF THE TOWN GLYPH (TUA-T)

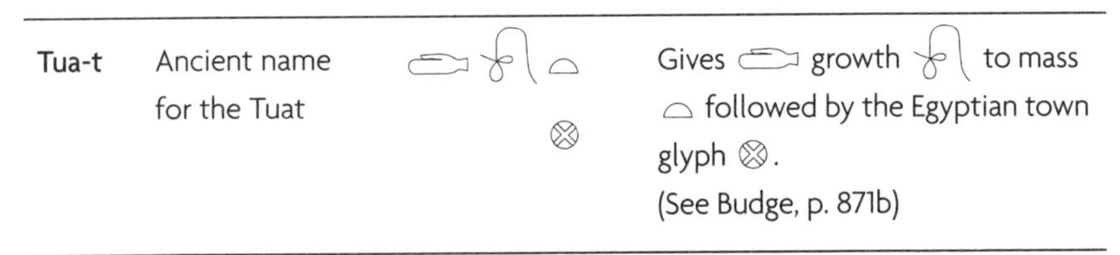

Tua-t	Ancient name for the Tuat		Gives growth to mass followed by the Egyptian town glyph . (See Budge, p. 871b)

The pylon or chamber glyph is another figure that we associate with the Tuat based on entries in our Structure of Matter table. Both the meaning of the pylon/chamber glyph and its relationship to the larger concept of the Tuat are established by two separate Egyptian defining word entries, each of which is pronounced "Tua-t." These two definitions, along with the preceding examples, show that the Egyptian hieroglyphic language consistently defines this same set of glyphs—the star glyph , the star in circle glyph , the Egyptian town glyph , and the pylon/chamber glyph —in terms of the hemisphere glyph , the notion of the growth of mass, and in relation to the Tuat itself.

DEFINITION OF THE PYLON/CHAMBER GLYPH (TUA-T)

Ancient name for the Tuat		Gives growth that becomes mass, followed by the Egyptian pylon/chamber glyph. (See Budge, p. 871b)

DEFINITION OF THE PYLON GLYPH (TUA-T)

Ancient name for the Tuat		Vibrations in the Calabi-Yau space become mass, followed by the pylon glyph. (See Budge, p. 871b)

The next question we should ask ourselves regarding these glyphs is whether any Egyptian words give us cause to associate the star glyph with the concept of vibrations. In *The Science of the Dogon,* we learned that Budge defines no Egyptian word for the concept of vibration. Rather, he chooses to translate various likely Egyptian references to vibration with a phrase of very similar meaning—the phrase "to tremble." The Egyptian icon used to convey the concept of trembling is the rabbit or hare glyph , which represents an animal that is known for its twitching and trembling. We can see from entries in the Structure of Matter table that aspects of the language itself lead us to associate the rabbit glyph with the spiraling coil glyph , which corresponds with the Dogon symbol of the coiled thread; both glyphs carry the same phonetic value of "un." We know that the sound "un" figures largely in the Egyptian concept of trembling, because the hieroglyphic word meaning "to tremble" is pronounced "unun" (see Budge, p. 167a). An Egyptian word with the same pronunciation, *unun,* also means "to do work in a field."[10] This definition brings us back full circle to the Dogon mythological field drawing whose image resembles the diagram in string theory that represents the vibration of a string. A related Egyptian word, *un-t,* meaning "rope or cord,"[11] constitutes a defining word for the spiraling coil or coiled thread glyph. It is written , or as we might interpret it ideographically, vibration creates mass, followed by the coiled thread glyph.

DEFINITION OF THE COILED THREAD GLYPH (UN-T)

Rope or cord		Vibration creates mass , followed by the coiled thread glyph . (See Budge, p. 167a)

It is not surprising that among these words relating to coiled threads and vibrations we also find an Egyptian defining word for the star glyph . Budge describes the word *unu-t* as meaning "time or hour." (It is an interesting side note that modern atomic clocks measure time very accurately based on the vibrations of particles of matter.) The glyphs of the word *unu-t* provide a definition for the star glyph, implying that the larger concept of time or hour—like the concept of the growth of mass—might refer to the notion of vibrations. The glyphs of the word *unu-t* read as follows:

DEFINITION OF THE STAR GLYPH (UNU-T)

Hour, time		Vibrations of particles of mass , followed by the star glyph . (See Budge, p. 167a)

Another Egyptian word that bears a relationship both to the root word *un* and the concept of vibrations is the word *unsh,*[12] which is a defining word for one of the jackal/wolf/fox/dog glyphs . Budge defines the meaning of this glyph as "wolf." If we recall the pattern set by other sequences of glyphs in the Structure of Matter table, such as the adze/sickle glyph sequence or the star/star-in-a-circle sequence, one key purpose of these sequences is to define the progressive steps of a process relating to the growth of matter. In this case, it is possible that the jackal/wolf/fox/dog symbols are meant to define a similar progression: as a mythological symbol, the jackal might represent mass in its disordered state, whereas the symbol of the wolf or fox may represent mass in a later, more organized state.

Each of the above references from the Egyptian language is consistent in the symbolism it assigns to this subset of glyphs, and they each support the conclusion that is suggested by our Structure of Matter table—that we should view the Tuat as an Egyptian counterpart to the Calabi-Yau space. By that interpretation, the star glyph would represent what Dogon mythology clearly tells us it represents—the vibrations of a string within the Calabi-Yau space. From that same perspective, the pylon/chamber glyph would represent one of the seven wrapped-up dimensions within the Calabi-Yau space. One Egyptian word for "chamber," *urit,*[13] is based on two other Egyptian phonetic roots, *ur* and *urr.* The first root, *ur,* carries the meaning of "wind," which is the mythological keyword for vibration; the second, *urr,* means "to increase, to grow."[14]

Budge defines another set of Egyptian words relating to the Tuat, based on the root *art,* or *arit.* He defines *arit* as "a judgment hall." This would seem like a very specific and appropriate type of chamber in which to house the judicial deliberations of the mythological wolf or jackal-god Sab. Budge also defines the word *Arit* as "a division of the Tuat" and tells us that the Arits were seven in number[15]—the same as the number of wrapped-up dimensions of the Calabi-Yau space defined by string theory. The word *Arit* is based on the phonetic root *ari,* which carries two definitions that are also directly related to the formation of matter. It first is defined as "wind," the mythological keyword for vibrations; second, it is described as "a kind of a fish," which once again calls to mind the nummo fish drawing of the Dogon that we interpret as representing the process of the perception and formation of mass. Another related Egyptian word is *ar-t,* which refers to the jawbone or the lower jaw and is a defining word for the Egyptian jaw glyph. This glyph takes the same general form as the sickle glyph but includes seven distinct teeth that are set into the jaw. The jaw glyph constitutes yet another glyph in the progression from the adze to the sickle to the jaw glyph, each symbolizing by its shape the progressive stages in the formation of mass.

Budge tells us that each of the seven Arits was in charge of a doorkeeper, a watcher, and a herald. The concept of a door in regard to the formation of mass was already described in the nummo fish drawing, in which it appears as the gateway to the twin Calabi-Yau spaces. So it is reasonable to presume that it could also represent the gateways between chambers or pylons within the Calabi-Yau space—the wrapped-up dimensions through which a string bends on its journey to create mass. One Egyptian word for doorkeeper is *arit-aa*. It is written as follows:

DEFINITION OF A DOORKEEPER (ARIT-AA)

Doorkeeper		That which bends mass , followed by the door glyph . (See Budge, p. 70a)

The Egyptian word for "watcher" is based on the root word *res*, which means "to watch" and seems closely related to the concept of perception—the initiating act by which mass is created. The word *res* is written and consists of symbols that by their very shape convey the notion of bending.[16] The word *resi* means "wind" (by our interpretation, "vibration") and is a defining word for the wind glyph ; the word *resres* means "to build."[17] So if the doorkeepers represent the passages between each of the wrapped-up dimensions of the Calabi-Yau space, then the watchers could easily be interpreted as representing the vibration and bending of the string that transpires within these dimensions.

One of the Egyptian words for herald is *uhemu,* meaning "teller, proclaimer, or herald."[18] This word is based on the root word *uhem*—found two entries higher on the same page in Budge's dictionary—which means "to repeat." Budge interprets another word *uhem* as meaning "to renew, to repeat an act."[19] Another word, *uhem-t,*[20] means "what is repeated, something that is renewed" and "a revolution of a star." However, we know that the mythological concept of a star refers to the seven windings or revolutions of a string as it passes through its repeated vibrations in the Calabi-Yau space. So we can interpret the Egyptian concept of herald as representing the repeated windings of the mythological space as it passes through the seven Arits of the Tuat.

The Dogon priests define a mythological Calabi-Yau space that is home to seven vibrations, the last of which "pierces the wall" of the space and passes to the outside to form an eighth "segment." These eight stages together are taken as representing the formation of the Word in Dogon mythology and seem to correspond to events that occur within the Tuat. If we are to believe that these symbolic interpretations are correct ones, then we would expect to find specific Egyptian references to tearing, cutting, or piercing related to the Egyptian concepts of the Tuat and the Word. In fact, upon examination we do encounter just such references. Phonetically, the first of these stands in direct relationship to the word *un* and the concept of vibration. From this root comes the name of the god Unti, who Budge defines as both "a light god, the god of an hour,"[21] and as "the opener, the piercer, the stabber."[22] The first of Budge's entries is a defining word for the star glyph and is written as follows:

DEFINITION OF THE STAR GLYPH (UNTI)

Light god, god of an hour	//	Vibration causes matter to be dual //, followed by the star glyph. (See Budge, p. 167b)

Budge's second word entry repeats the same ideographic message and is a defining word for a glyph that symbolizes the concepts of piercing and opening.

DEFINITION OF THE PIERCING/STABBING GLYPH (UNTI)

Opener, piercer, stabber		Vibration causes matter to be dual //, followed by the piercing/stabbing glyph . (See Budge, p. 166b)

Yet another obvious reference to the notion of cutting is found in the name of the Egyptian god Tua, who Budge defines as the "god of circumcision."[23] Circumcision is, of course, the ritual cutting of the foreskin that, in Hebrew tradition, occurs on the eighth day after the birth of a male child. Another Egyptian word, *tua,* refers to a "knife, scalpel, or knife used in circumcision."[24] At the heart of these two words is the phonetic root "tu," which we mentioned previously. Budge defines an Egyptian word *tu* as meaning "bandlet."[25] Based on our stated criteria, this word can be seen as a defining word for the looped string intersection glyph .

DEFINITION OF THE LOOP STRING INTERSECTION GLYPH (TU)

Bandlet		Made from the coiled thread , followed by the looped string intersection glyph . (See Budge, p. 870a)

Another link to the concept of cutting or dividing is found in the family of words formed from the phonetic value "tch." This is the phonetic value Budge assigns to the serpent glyph and is the very glyph within the Egyptian hieroglyphic language that symbolizes the Word. Budge defines the word *Tchet* as meaning the "Divine Word." It is written:

DEFINITION OF THE FALCON GLYPH (TCHET)

The Divine Word		Serpent given growth, followed by the falcon glyph/god determinative. (See Budge, p. 913b)

This word is closely related to the word *tches,* which carries two meanings, both of which have significance for our inquiry; according to Budge it can mean "to cut or divide,"[26] and it also can mean "knife."[27] Likewise, there is a related word, *tchet-t,* that means "the culmination of a star," a definition that complements the symbolic star imagery of the Tuat. Another Egyptian word, *tchet,* means "permanent, abiding, enduring, established." We might interpret this as a reference to the completion of a membrane—an enduring subcomponent of matter in string theory. Budge also defines the word *Tchet-s* as "a serpent goddess who was reborn daily." This meaning repeats the time/hour/day imagery that we discussed earlier in regard to the Tuat and combines it with the serpent imagery found at the Tuat's completion. (The Tuat as counterpart to the Calabi-Yau space constitutes a progression that comes to completion. In Egypt, the Tuat is associated with the hours of the night, which also come to completion.) The Tuat is associated with dawn or the beginning of day. Words that define the vibrations within the Calabi-Yau space are associated with the marking of time or hours. The completed Calabi-Yau space—the Divine Word—is alternately defined as a serpent goddess who is reborn daily.

Each of these references to time make some sense from an interpreted view of science because, prior to the formation of mass, it appears that unperceived waves exist in a nearly infinite time frame, a context within which the concept of time essentially would be meaningless. The act of perception initiates the successive vibrations within the Calabi-Yau space that slow the time frame of the wave, endow it with mass, and cause it to become subject to the effects of gravity. We see strong hints of this perspective in Egyptian words relating to the concept of the hour. One Egyptian word for hour is *unu-t.* In perhaps its simplest form, the word *unu-t* is written as follows:

DEFINITION OF THE GRAVITY GLYPH (UNU-T)

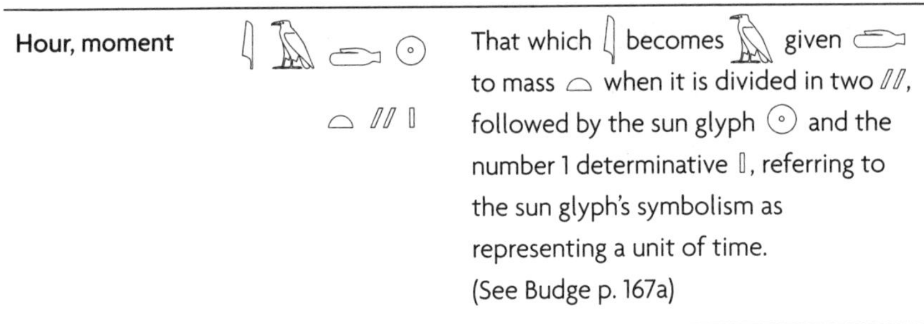

Hour, moment		That which becomes given to mass when it is divided in two //, followed by the sun glyph and the number 1 determinative, referring to the sun glyph's symbolism as representing a unit of time. (See Budge p. 167a)

Another word for "hour" is *aat-t;* its phonetic root word is *aat,* which means "speech." The word *aat-t* is written as follows:

DEFINITION OF THE SUN GLYPH (AAT-T)

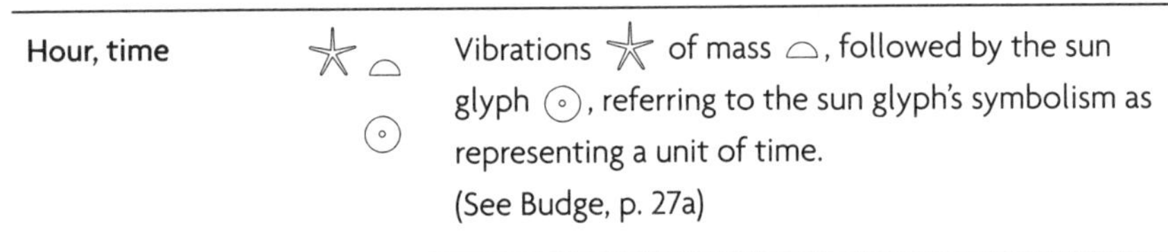

Hour, time		Vibrations of mass, followed by the sun glyph, referring to the sun glyph's symbolism as representing a unit of time. (See Budge, p. 27a)

From these varied examples, we see that the key Egyptian concepts involving the Tuat closely follow those of string theory and the Calabi-Yau space as well as with Dogon mythological references regarding the vibrations of the mythological cosmic thread. Such parallels follow us through level after level of language and symbolism, matching mythological concepts with scientific concepts and Dogon words with Egyptian words, all in close association with the Tuat. These parallels are borne out both by the common-usage meanings and ideographic definitions of the Egyptian words and by the very shapes of the glyphs used to write these words. The suggestion is that the jackal represents mass in its disordered state and that the wolf, fox, or dog represents matter in its reordered state; both states of matter appear to fall under the domain of the Tuat. This conclusion is supported by another set of Egyptian words that relate to the formation of membranes at the completion of the Calabi-Yau space—words that are based on the Egyptian root words *a,* which means "to roll," and *aa,* which means "spacious." *Auau* is the Egyptian word for dog or jackal (see Budge, p. 4a) and refers to the sound a dog makes as it opens its mouth—an act that can be seen as a metaphor for the creation of space or spaciousness based on the V shape formed by the dog's open mouth. A related Egyptian word, *aa-t,* means "pale"[28]—the very description from Dogon mythology that is associated with the creation of mass in *The Pale Fox.*

ELEVEN

EGYPTIAN PHONETIC VALUES

During our previous discussions of the Egyptian language and the structure of matter, we have come to see that the phonetic value of a glyph can be an important key for unlocking its symbolism. Ultimately, it was the phonetic values of individual glyphs that led us to the names of corresponding deities and to the supporting words that comprise our Structure of Matter table. This methodology runs contrary to the initial approach we took when assigning symbolism to Egyptian glyphs, which had little to do with phonetic values; rather, it was based primarily on the close resemblances of Dogon mythological drawings to specific Egyptian glyph shapes. In fact, because this symbolism was initially assigned to many of the glyphs without real regard to phonetics, our method should offer little reason to suspect that a relationship might exist between the phonetic values of glyphs and concepts of science. However, as we explore Egyptian phonetics in greater depth, we will begin to see that such a relationship might well exist.

To understand the possible relationship between phonetic values and scientific meaning, we should begin by considering the subset of Egyptian glyphs that carry direct phonetic values. The majority of these glyphs are easily found by browsing through Budge's List of Hieroglyphic Characters and selecting those with an entry in the phonetic value column. Others can be found by looking to single-glyph word entries within the body of Budge's dictionary. Over time, modern Egyptologists have come to disagree with Budge's assignment of phonetic values to specific glyphs, and we should emphasize that it is *not* our purpose here to try to rehabilitate Budge's phonetic assignments. However, so long as Budge applies his values in a consistent way, it essentially is irrelevant to our argument whether a given glyph should actually be pronounced "a" instead of "e" or "x" instead of "kh." During the course of this discussion, we should consider Budge's phonetic values as tentative placeholders, which we will use in association with various glyphs to illustrate a specific set of relationships; later, if we choose to substitute a different set of phonetic values for Budge's, we should see that this set of relationships still holds true.

As we begin these discussions, perhaps our first step should be to develop a list of glyphs that carry direct phonetic values so that we can consider them in the context of a discrete group. As we compile this list, we should make note of any hieroglyphic characters Budge uses to clarify the meaning or phonetic value of a glyph; later, these symbols—which appear in the fourth column of the table that follows—may become important to our discussions. Along the way, if we encounter a glyph whose shape and phonetic value have played no role in our discussions of the structure of matter, we may choose to omit the glyph from our list for the sake of brevity. Likewise, as we select individual glyphs for inclusion in the list, it makes sense to organize the glyphs by phonetic value. After this selection process has been completed successfully, we find that Budge's phonetic values and glyph definitions can be used to produce the table on the next page.

EGYPTIAN GLYPHS WITH DIRECT PHONETIC VALUES

Glyph	Phonetic Value	Budge's Definition and Page Reference*	Tentative Ideographic Definition†
	a	None (p. cxxii)	
	a	I, me, my (pronominal suffix) (p. 15a)	
	a[1]	God or divine person (p. ci)	
	a	An emphatic particle (p. 1a)	
	a	Piece, one, a, an, pair (p. 105a)	
	a	Forearm, hand (p. 105a)	
	aa	Great (p. cxxxviii)	"Act or force comes to be"
	aha	Stand up (p. cxxx)	
	am	Side (p. cxxvi)	"That which is beheld/known"
	an	Fish (p. cxx)	"That which weaves"
	antch	Sound, healthy (p. cxli)	"The force that weaves the serpent"
	ar	Right eye, to see (p. cv)	"That which is bent/warped" "The act of weaving"

*From *An Egyptian Hieroglyphic Dictionary.*

†Interpreted from Budge's explanatory glyphs.

1. These glyphs carry no phonetic value when used as a determinative.

Glyph	Phonetic Value	Budge's Definition and Page Reference	Tentative Ideographic Definition
	ar	Pupil of the eye (p. cvi)	"That which is bent/warped"
	ari	Present at, belonging to (p. cii)	// "That which is bent to or toward"
	au	Go, pass, like, similar[2] (p. cxxvi)	"That which is grown"
	b	Abode, place (p. 197a)	"Place grows"
	e[3]	None (p. cxxii)	
	e	I, me, my (pronominal suffix) (p. 15a)	
	f	Horned viper (p. cxix)	
	f	He, his, its (p. 258a)	
	h	Shelter, windbreak (p. 438a)	(In Nubian texts for)
	i[4]	None (p. cxxii)	
	i	I, me, my (pronominal suffix) (p. 15a)	
	i	None (p. cxlvi)	//
	i	An exclamation (p. 142a)	
	k	None (p. cxliv)	
	k	Thou, thee, another, also (p. 782a)	"Collected together" "Collected together body"

2. Later researchers interpret this glyph to mean "island."

3. The assignment of "e" is problematic for later Egyptologists.

4. Later researchers interpret this glyph as "y" or "a."

EGYPTIAN GLYPHS WITH DIRECT PHONETIC VALUES *(cont.)*

Glyph	Phonetic Value	Budge's Definition and Page Reference	Tentative Ideographic Definition
	ka	The double (p. cvii)	
	kes	Bow, pay homage (p. xcviii)	"Collected mass bends/binds"
	kh	Sieve (p. cxlv)	
	kh	None (p. 525a)	
	kha	Part of papyrus plant (p. cxxiii)	"Result comes to be"
	kha[5]	None (p. cxxxviii)	
	kha	None (p. 570a)	
	khep[6]	He creates what is (p. 541a)	"Source of space"
	kher	Fall, defeat (p. cii)	"Effect of gravity"
	khet	Tree, wood (p. cxxi)	"Source of mass"
	khu	Rule, govern (p. cvii)	"Source of growth"
	m	None (p. cxv)	
	m	In, into, from, on, at, with, out (p. 264a)	
	m	Mark of negation (p. 266a)	
	m	See, behold (p. 266a)	

5. Later researchers interpret this glyph as a soft "ch."

6. The full value of this sign is "kheper." The final "er" is lost in some cases.

Glyph	Phonetic Value	Budge's Definition and Page Reference	Tentative Ideographic Definition
	m	Come (p. 266b)	
	m	Give (imperative only) (p. cvii)	[7] "Bending/ warping given"
	m	Side (p. cxxvi)	"That which is beheld/known"
	m	Mark of negation (imperative only) (p. 266a)	
	m	See, behold (p. 266a)	
	m	Come (p. 266b)	
	maa	True, right, truth, integrity (p. cxxxiii)	"The perceiving force"
	maa	To examine, to perceive, to inspect (p. 266b)	
	men	Draughtboard (p. cxlvi)	"Completely woven"
	mu	Water, watery mass of the sky (p. cxxvi)	
	n	None (p. cxxvi)	
	n	While, as long as, because, since, for, then (p. 339a)	
	n	Lack, want, need, nothing, no, not (p. cvii)	
	nes	Tongue, leader[8] (p. cxiii)	, "Waves bind/waves bend"

7. The above two glyphs from Budge are actually pronounced "rdi."
8. When used to mean "leader," this glyph is not pronounced "nes."

EGYPTIAN GLYPHS WITH DIRECT PHONETIC VALUES *(cont.)*

Glyph	Phonetic Value	Budge's Definition and Page Reference	Tentative Ideographic Definition
	nu	Blade of an adze (p. cxxxix)	"Waves grow"
	nu	Vase, vessel, pot (p. cxliii)	
	p	Door (?)[9] (p. cxxix)	
	p	My, mine[10] (p. 229a)	
	pa	Duck, waterfowl, flying (p. cxvi)	"Space comes to be"
	q	None (p. cix)	
	q	None (p. 760a)	
	qa	High, lofty (p. xcvii)	"Earth comes to be"
	qes	Restrain, bind (p. xcix)	"Earth bound"
	qes	To tie, bind, cordage (p. cxli)	"Earth bent"
	qet	To build (p. xcix)	
	qet	None (p. cxxix)	"Earth given"
	r, ra	Mouth (p. cvi)	
	r	At, by, near, to, toward, into, with (p. 414a)	"That which bends or warps"
	res	The South (p. cxxii)	"Warp or bend"

9. Budge is uncertain about the assignment of the meaning "door" to this glyph.

10. *P* only means "my" or "mine" as an abbreviation of a larger word.

Glyph	Phonetic Value	Budge's Definition and Page Reference	Tentative Ideographic Definition
	rmn	To arm, bear, carry (p. cvii)	"Bend, draw, or weave"
	s	None (p. cxxxi)	
	s	She, he (p. 633a)	"Bending grows"
	s	She (p. 583a)	
	sa	Goose; the Earth god Geb (p. cxvi)	"Bending comes to be"
	sba	Star, morning star, hour, time (p. cxxv)	"Bending place comes to be"
	sen	Like, similar (p. cxxvi)	"Bound and woven"
	ser	Great, great one (high official) (p. xcviii)	"Bound together"
	set	Underworld (p. cxi)	
	shes	Tie, bind, cordage (p. cxli)	
	su	Plant of the South (p. cxxii)	"Bend and grow"
	t	None (p. cxlvii)	
	t	Thou, thee (p. 815a)	"Mass"
	t	Hand, palm of the hand (p. cviii)	"Serpent becomes matter"
		To give, to set, to place (p. 864)	

EGYPTIAN GLYPHS WITH DIRECT PHONETIC VALUES *(cont.)*

Glyph	Phonetic Value	Budge's Definition and Page Reference	Tentative Ideographic Definition
	ta	None (p. cxxxix)	"Mass," "mass exists"
	t	Whilst, when, as, because, to give (p. 815a)	"Mass given"
	ta	Land (p. cxxv)	"Mass comes to be"
	tch	Serpent (p. cxix)	
	tch	Serpent that came forth from Ra (p. 893a)	
	tchu	Mountain (p. cxxv)	"Serpent grows"
	tchet	Sacred object (p. cxxxiii)	"Serpent given"
	tem	Finish, complete, bring to an end (p. cxxxix)	"Mass beheld," "mass complete"
	th	Thee, thou (p. 848a)	
	tu	Mountain (see "tchu" above) (p. cxxv)	"Gives growth"
	tua	Star, morning star, hour, time (p. cxxv)	"Given growth comes to be"
	tua	Pray, worship, adore (p. xcvii)	"Vibration and growth come to be" Synonym for "That which becomes grown"

Glyph	Phonetic Value	Budge's Definition and Page Reference	Tentative Ideographic Definition
	u	Chicken, quail (p. cxvii)	
	u	They, them, their (p. 144a)	
	u	Cord (p. cxxxiv)	"Growth"
	ua	None (p. cxlii)	"Growth comes to be"
	ua	Pike, harpoon (p. cxl)	"Growth force"
	uab	Pour out water (p. ci)	"Growth force of place"
	uatch	Papyrus stalk (p. cxxii)	, "Growth of the serpent"
	up	Crown of the head, apex (p. cxiii)	
	un	Hare (p. cxi)	
	un	None (p. cxxiii)	"Growth of waves" "Growth of the coiled thread"
	ur	Great, great one (p. xcviii)	"Growth of gravity"
	ur	Swallow (p. cxvii)	

As we consider this list of glyphs, perhaps the most obvious point to notice is that the assigned phonetic values of the glyphs take two basic forms. When transliterated into English, some can be represented as single-letter sounds, such as "n" ( ), "p" (□), "r" ( ), and "t" ( ), and others require two or more combined English letters to form their sound, such as "aha" ( ), "khet" ( ), "tua" ( ), and "tch" ( ). Because our table is organized phonetically, glyphs that begin with similar phonetic values fall together in the table, and so in many cases a glyph with a single-letter phonetic value immediately precedes one with a multiple-letter value. A simple example of this would be the square glyph (□), which carries the phonetic value of "p," followed by the flying goose glyph ( ), which Budge pronounces "pa." In our discussions of the structure of matter, we assigned the scientific concept of space to the square glyph. We notice in our table that Budge defines the flying goose glyph with two other glyphs (□), which we can read ideographically as "space comes to be," an appropriate definition for a symbol we have come to associate with the atom. Another similar example involves the ( ) glyph, which Budge pronounces "q." This is a character that we interpret ideographically as symbolizing earth or mass. It is followed in our table by the ( ) glyph, which carries the phonetic value of "qa." As in the previous case, Budge includes a two-glyph phonetic definition for this glyph ( ), which we read ideographically as "earth comes to be." Likewise, in our Structure of Matter table, the word *qa* and the glyph define the concept of earth (or mass) raised up, or earth coming to be.

Now, if we ignore the glyphs themselves and focus solely on the relationship of the phonetic values involved, we can see that the value "a," when attached either to the "p" or the "q" sound, seems to imply the concept of coming to be or coming into existence. Likewise, Budge's referential glyphs show that he specifically relates the "p" sound in the flying goose glyph to the "p" sound in the square glyph, as does the "q" sound in the next example relate to the "q" sound in the earth glyph that precedes it. What this suggests is that when the phonetic value of a glyph is transliterated to multiple-letter sounds, each component sound actually may refer to a glyph with a single-letter phonetic value, and thereby to the assigned concept associated with that glyph. These two simple examples show that if we join the concepts associated with these single-letter glyphs together, we can produce a combined concept that defines the multiple-letter glyph. If this pattern were to hold true for all glyphs, and if we could establish a working list of concepts to associate with each single-letter phonetic value, then we might find that we are able to predict which ideographic concept to associate with each multiple-letter glyph based solely on its phonetic value.

Based on previous discussions in both *The Science of the Dogon* and *Sacred Symbols of the Dogon,* we already have assigned tentative concepts to most of the glyphs that carry single-letter phonetic values, so we can arrive at an initial list of glyphs and concepts simply by adopting these as base values. For situations in which we find that more than one glyph carries the same phonetic value, our reasonable assumption might be that each glyph correlates to its own distinct meaning or concept. In other cases, such as the reed leaf glyph ( ), we find multiple phonetic values ("a," and "i" or "y"; Budge also infers a disputed value of "e") associated with a single glyph. Again, the reasonable assumption might be that each distinct pronunciation corresponds to its own concept.

In the Egyptian hieroglyphs—as in English—vowels can carry different pronunciations in different words. For instance, in English, we pronounce a short "a" in the word "ma" and a long "a" in the word "May." Even though regional dialects result in differences in pronunciation that often can be more subtle than the simple case of long and short sounds, in English, we consistently acknowledge that an "a" is an "a." So the question arises as we assign ideographic meaning to Egyptian glyphs: Should the short and long "a" sounds and the dotted "a" sound imply distinct and separate ideographic concepts, or do the various "a" sounds belong together conceptually? For an initial answer to this question, we might consider the previously discussed mind-set of the authority under which this symbolism presumably evolved. Based on the pattern exhibited within Dogon mythology—a pattern that seems to be consistently mirrored by Egyptian glyphs and words—we infer that it was *not* the intention of that authority to complicate an issue unnecessarily, even in regard to the definition of seemingly complex issues. Likewise, the case can be made based on references within the Egyptian language itself for considering a dotted "a" as the conceptual sister of the other "a" sounds. For instance, the word *An-t,*[1] which is transliterated by Egyptologists with a dotted "a" sound, refers to a mythological fish; so does the word *an,*[2] which is transliterated with a long "a" sound. The same is true of the words *ua,* transliterated with the dotted "a" sound,[3] and *ua,* transliterated with the long "a" sound,[4] both of which refer to a kind of fish. Based both on the general mind-set of the mythology and these kinds of specific case examples, the simple answer may be that, until otherwise indicated, we should consider all ideographic "a" sounds to be of a kind. Based on these assumptions and other clues of context, we can establish the following tentative list of glyphs, phonetic values, and concepts found on page 159.

If it is true that each phonetic value represents a concept, then the suggestion is that we might mix and match phonetic values to predict which concept will be represented by other compound phonetic glyphs. For instance, if "t" represents "mass," and "m" represents "complete," then the glyph that is pronounced "tem" (the "e" vowel sound is merely inferred for convenience) should represent the concept of mass complete. We can test this proposition by looking through our table titled Egyptian Glyphs with Direct Phonetic Values and finding the glyph pronounced "tem" ( ). We see that Budge assigns two referential glyphs ( ) to his entry, which based on our system of interpretation, carry the ideographic meaning of "mass observed" or "mass complete." Likewise, if "q" stands for "earth" and "t" stands for "given," then the glyph that

EGYPTIAN PHONETIC VALUES AND CONCEPTS

Glyph	Value	Concept
	a	To be, to create, to exist, to become, to come into existence
	b	Place
	e	Is, of, and (note: "e" is a disputed value for this glyph)
	f	To transmit
	h	Structure, courtyard, to rotate (see the words "hett" and "hethet," Budge, p. 452a)
	i	Duality, together, by, origin (that which is that which is)
	k	To collect together, to combine, collected body (quark)
	m	To see, behold or know, complete, to give, side, mark of negation
	n	Wave, to weave, to waver, to write, to create, to cause
	p	Space
	q	Earth, mass
	r	The bending/warping of gravity
	s	To bend or bind (in regard to bodies of mass)
	t	Mass
	u	Growth, vibration

is pronounced "qet" should carry an ideographic definition similar to "earth given." Again, we search our table of glyph phonetic values and locate the glyph with the value of "qet" ( ), and we see that Budge has included two referential glyphs ( ), which we interpret ideographically to mean "earth given."

We can apply this same technique to another set of Egyptian glyphs, beginning with the wave glyph ( )—pronounced "n"—to which we assign the concept of a massless wave. This glyph is followed a few entries later in our table by the adze glyph ( ), which Budge pronounces "nu." Based on our assigned phonetic values, "n" stands for "waves" and "u" for "growth," so the combined value "nu" should refer to "wave growth." Once again, Budge obligingly provides us with two defining glyphs for the adze glyph ( ), which we read ideographically as "wave grows."

As we browse through the Egyptian Glyphs with Direct Phonetic Values table, we find many examples of glyphs that fit this same basic pattern. For instance, the phonetic value "ka" ( ) brings us to the glyph we associate with a pair of quarks. In the structure of matter, a quark is perhaps the earliest collected body of mass. Based on our phonetic value table, "k" represents "to collect together" or "collected body" and "a" represents "to exist," so the combined concept would read "collected body exists." Likewise, if the phonetic value "s" (or) means "to bend or bind," then the phonetic value "kes" ( ) would read "collected body bound." This is an appropriate meaning for a glyph that we have interpreted as representing the weak nuclear force, which weakly binds quarks into protons and neutrons.

Another good example of this same pattern is found with the glyphs that carry the phonetic value "qes" ( ). According to our assigned phonetic concepts, "q" represents "earth" and "s" represents "to bend or bind," so the combined value "qes" should imply the concept of "earth bent or bound." Budge's defining glyphs for the glyph are , which can be read ideographically as "earth bent"; his entries for the glyph are , or "earth bound."

One of the more interesting glyphs from a phonetic standpoint is what Budge calls the sieve glyph ( ), which to our way of thinking can also depict a pool of water. This glyph is pronounced "kh," and based on our phonetic concepts would mean "collected structure," much as a lake is a collected pool of water. Interpreted from this viewpoint, a collected structure might imply several different meanings, which we have already discussed in regard to this glyph and supported with specific case examples. It can suggest the concept of source, in the way that a lake can be seen as the collected body that is the source of a river. It also can suggest a limit, much as the surrounding bank of a lake is the limit of the collected pool of water it contains. In addition, it can also suggest the end product, result, or effect of a process, in the same sense that a pooled body of water is the result, end product, or effect of collected rainfall. Based on these definitions, if in some cases we take the combined phonetic value of "kh" to mean "source," then the phonetic value "khep" ( ) would refer to "source [of] space," and the value "khet" ( ) would refer to "source [of] mass" (again, the mythological reference is to wood, which is what fire burns). In other cases if we take the phonetic value "kh" to mean "effect," then Budge's referential glyphs ( ) for the falling man glyph ( ), which is pronounced "kher," make sense because "kh" represents "effect" and "r" represents "gravity," or "effect [of] gravity."

Other specific glyph examples make similar good sense when interpreted in the context of our table of phonetic values. For instance, the phonetic value "set" ( ), which combines the "s" and the "t" phonetic values, would imply a combination of the ideographic concepts represented by those sounds, or "bending [of] mass," the same concept that is implied by supporting references in our Structure of Matter table. Likewise, the phonetic value "tua," which we believe refers to the Calabi-Yau space, is associated with two separate glyphs (and), and would imply the concept "mass vibrations come to be."

Following the pattern set by these examples, we could apply the same logic to each of the other Egyptian glyph phonetic values and interpret a meaning for the glyph based purely on the components of its phonetic value as Budge pronounces it. Such an approach results in the following table of values:

INTERPRETED GLYPH PHONETIC VALUES*

Glyph	Phonetic Value and Budge Page Reference†	Phonetic Interpretation
	a (p. cxxii)	To be, to exist, that which is, that which comes into existence
	a (p. 15a)	I, me, my (pronominal suffix)
	a (p. ci)	God or divine person (determinative)
	a (p. 1a)	To become, to come to be
	a (p. 105a)	Piece, one, a, an, pair
	a (p. 105a)	Forearm, hand (ideographically, act or force)
	aa (p. cxxxviii)	Act or force comes to be, act or force applied
	aha (p. cxxx)	Creative structure exists
	am (p. cxxvi)	That which is beheld (perceived, conceived)
	an (p. cxx)	That which weaves
	antch (p. cxli)	That which weaves the serpent
	ar (p. cv)	That which bends or warps (perception)
	ar (p. cvi)	That which is bent or warped (an orbit)
	ari (p. cii)	That which is bent to or toward
	au (p. cxxvi)	That which is grown (symbol of protons, neutrons, and electrons)

*Many of the single-glyph word values that Budge assigns in the body of his dictionary are also listed.

†From *An Egyptian Hieroglyphic Dictionary*.

Glyph	Phonetic Value and Budge Page Reference	Phonetic Interpretation
	b (p. 197a)	Place
	e (p. cxxii)	Is, of, and (the assignment of "e" to this glyph is problematic)
	e (p. 15a)	I, me, my (pronominal suffix)
	f (p. cxix)	To transmit?
	f (p. 258a)	He, his, its
	h (p. 438a)	Structure
	i (p. cxxii)	To, toward
	i (p. 15a)	I, me, my (pronominal suffix)
	i (p. cxlvi)	Dual, duality, together, by
	i (p. 142a)	An exclamation
	k (p. cxliv)	Collect together, combine, collected body
	k (p. 782a)	Thou, thee, another, also
	ka (p. cvii)	Collected body exists (quark)
	kes (p. xcviii)	Collected body bent/bound (weak nuclear force)
	kh (p. cxlv)	Collected structure (source, limit, result, or effect of)
	kha (p. cxxiii)	Collected structure comes to be (increased mass)
	kha (p. cxxxviii)	Source of existence (womb)
	kha (p. 570a)	Source of existence (womb)
	khep (p. 541a)	Source of space

INTERPRETED GLYPH PHONETIC VALUES *(cont.)*

Glyph	Phonetic Value and Budge Page Reference	Phonetic Interpretation
	kher (p. cii)	Effect of gravity (falling down)
	khet (p. cxxi)	Source of mass (fire burns wood)
	khu (p. cvii)	Source of growth
	m (p. cxv)	To see or perceive, to behold or know, complete
	m (p. 264a)	In, into, from, on, at, with, out
	m (p. 266a)	Mark of negation
	m (p. 266a)	See, behold
	m (p. 266b)	Come
	m (p. cvii)	To give
	m (p. cxxvi)	Complete, beheld, known (concept)
	maa (p. cxxxiii)	Perception comes to be
	maa (p. 266b)	Behold that which is (perception)
	men (p. cxlvi)	Complete and woven
	mu (p. cxxvi)	Perception grows (waves)
	n (p. cxxvi)	Wave, to weave, to write
	n (p. 339a)	While, as long as, because, since, for, then
	n (p. cvii)	Lack, want, need, nothing, no, not, nonexistence
	nes (p. cxiii)	Waves are bent/bound
	nu (p. cxxxix)	Waves grow
	nu (p. cxliii)	Waves grown

Glyph	Phonetic Value and Budge Page Reference	Phonetic Interpretation
	p (p. cxxix)	Space
	p (p. 229a)	My, mine
	pa (p. cxvi)	Space exists (atom)
	q (p. cix)	Earth
	q (p. 760a)	Earth
	qa (p. xcvii)	Earth exists
	qes (p. xcix)	Earth is bent/bound
	qes (p. cxli)	Earth is bent/bound
	qet (p. xcix)	Earth and mass
	qet (p. cxxix)	Earth is given
	r, ra (p. cvi)	Gravity, gravity exists
	r (p. 414a)	At, by, near, to, toward, into, with
	res (p. cxxii)	Gravity is bent
	s (p. cxxxi)	To bend or bind
	s (p. 633a)	She, he
	s (p. 583a)	To bind
	sa (p. cxvi)	Bending comes to be
	sba (p. cxxv)	Bending place exists
	sen (p. cxxvi)	Bound and woven
	ser (p. xcviii)	Binding of gravity (strong nuclear force)

INTERPRETED GLYPH PHONETIC VALUES *(cont.)*

Glyph	Phonetic Value and Budge Page Reference	Phonetic Interpretation
	set (p. cxi)	Bending of mass
	shes (p. cxli)	Bent structure is bound
	su (p. cxxii)	Bending grows
	t (p. cxlvii)	Mass
	t (p. 815a)	Thou, thee
	t (p. cviii)	To give
	t (p. 864)	To give, to set, to place
	t, ta (p. cxxxix)	Mass, mass exists
	t (p. 815a)	Mass given
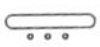	ta (p. cxxv)	Mass comes to be
	tch (p. cxix)	Mass becomes structured (serpent or Word)
	tch (p. 893a)	Serpent that came forth from Ra
	tchu (p. cxxv)	Serpent grows
	tchet (p. cxxxiii)	Word is given
	tem (p. cxxxix)	Mass is complete
	tet (p. cvii)	Mass is given (force)
	th (p. 848a)	Mass structured
	tu (p. cxxv)	Mass given growth/shape
	tua (p. cxxv)	Mass given growth and existence
	tua (p. xcvii)	Mass given growth exists

Glyph	Phonetic Value and Budge Page Reference	Phonetic Interpretation
	u (p. cxvii)	Grows, growth
	u (p. 144a)	They, them, their
	u (p. cxxxiv)	Coiled thread, growth
	ua (p. cxlii)	Growth exists
	ua (p. cxl)	Growth exists
	uab (p. ci)	Growth of place
	uatch (p. cxxii)	Growth of the serpent/Word
	up (p. cxiii)	Growth of place
	un (p. cxi)	Growth of waves (vibration)
	un (p. cxxiii)	Grown waves (electron)
	ur (p. xcviii)	Growth of gravity
	ur (p. cxvii)	Growth of gravity

The preceding discussion implies that there may well be an underlying logic to the assignment of phonetic values to Egyptian glyphs, one that is both consistent and predictive. If that is true, then these values provide us with yet another avenue of confirmation for the scientific meanings of these glyphs, along with examinations of their ideographic forms, Dogon drawings, the roles of corresponding deities, and the meanings of similarly pronounced Egyptian words. In short, it would seem that all signs point in the same direction—to a system of scientific symbolism based on the phonetic values of Egyptian glyphs.

TWELVE

REVISITING THE SYMBOLISM OF DOGON COSMOLOGY

We began this study by using elements of Dogon cosmology and language to clarify the meanings of various Egyptian glyphs and words. This process has effectively revealed an Egyptian structure of matter—couched in the symbols of language—that at times seems to reflect a higher degree of symbolic detail than the Dogon model that led us to it. This implies that the Egyptian hieroglyphs may offer us important insights into related Dogon words and symbols and suggests that we should now turn our gaze back in the direction from which we came to see if we can further clarify any of the original statements of Dogon cosmology based on these new insights. Likewise, our study of the Egyptian hieroglyphs has provided us with a new framework for organizing and comparing these parallel words and symbols—the framework of matter itself as it is reflected in our Structure of Matter table. If we use that table as a guide, the logical place to start our comparisons would be at the beginning—at the point where existence emerges from nonexistence.

Nonexistence

Egyptian concepts of nonexistence and coming into existence are defined by the word *kheper.* Words with this pronunciation take the form of defining words for the dung beetle or scarab glyph and form the basis of the word *kheprer,* an Egyptian word for beetle and the name of an Egyptian beetle god.[1] Genevieve Calame-Griaule tells us in *Dictionnaire Dogon* that the corresponding Dogon prefix *ke* refers to the dung beetle specifically and to all water beetles in general and so refers to insects that emerge from water. The Egyptian word *kheper,* meaning "nonexistence," is written as follows:

DEFINITION OF THE SCARAB GLYPH IN ITS ASPECT AS NONEXISTENCE (KHEPER)

Nonexistence		Not created , followed by the scarab glyph . (See Budge, p. 542a)

We see that the Egyptian concept of nonexistence is defined ideographically in terms of the wave glyph and therefore in terms of water. Budge lists another spelling of the same word *kheper,* meaning "to be, to exist, to come into being, to fashion," which qualifies as an alternate defining word for the same scarab glyph. This definition is expressed in terms of a collected pool and therefore also in terms of water.

DEFINITION OF THE SCARAB GLYPH IN ITS ASPECT OF COMING INTO EXISTENCE (KHEPER)

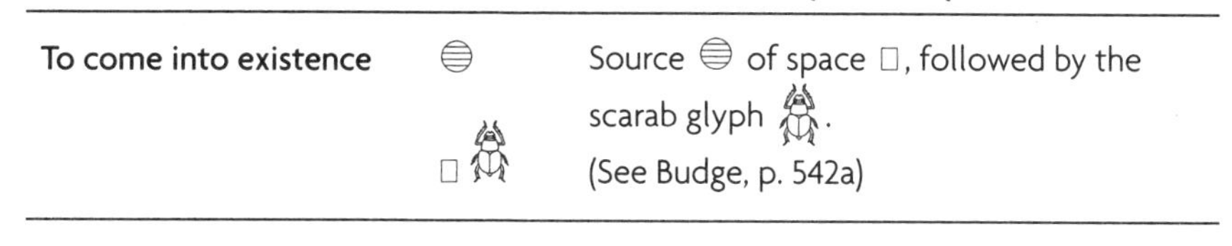

To come into existence		Source of space , followed by the scarab glyph . (See Budge, p. 542a)

Creation from Water

The theme of creation from water is one that is central to both Dogon and Egyptian mythology. This aspect of creation is defined in both the Egyptian hieroglyphic language and Dogon cosmology by the sound "nu." The phonetic value "nu" forms the root of the Dogon word *nummo,* which the Dogon define as the perfect twin pair that emerges at the time of creation; the Dogon priests affirm that the word *nummo* specifically refers to water (see *Conversations with Ogotemmeli*). An ideographic reading of the Egyptian word *nu* reflects this same symbolism. Budge defines the word *nu* as meaning "mass of water that existed in primeval times" and "deified primeval waters whence everything came."[2] In broad terms, we can say that Egyptian hieroglyphic words affirm that the term *nu* refers to primeval waters of creation and that Dogon cosmology reaffirms that the concept relates to the formation of the universe and of matter. In a note to one of his dictionary entries, Budge states that the word *nu* is considered the plural of the wave glyph,[3] which carries the phonetic value of "n."

WAVES GROW (NU)

Plural of the wave glyph	[wave glyph]	Waves [wave glyph] grow [bird glyph].
	[bird glyph]	(See Budge, p. 349a)

Perception

The concept of perception—the act that initiates the formation of matter—is directly linked to the concept of nu by way of the Dogon word *nummo*. We interpret *nummo* as the Dogon counterpart to the Egyptian phrase *nu maa*. We know that the Egyptian word *nu* refers to the "primeval waters," and Budge tells us that *maa* means "to examine or perceive."[4] Likewise, the Egyptian sickle glyph [sickle glyph], which we have interpreted as a symbol for perception, is pronounced "maa." We know that the word *maa* in its simplest form is written with the single sickle glyph, and so the sickle becomes a symbol for perception. Based on these definitions and observations, we can interpret the combined term *nu maa* as meaning "waves perceived."

For the Dogon, this initiating act of perception is a complex process, and for our purposes is one that requires extended discussion. Based on our earlier interpretations, the process of the perception of matter is defined in stages by elements of the nummo fish drawing. We look first at the egg in a ball, the spiked ball figure that we identified as the point of perception at the center of the fish. In *The Pale Fox,* Griaule and Dieterlen present a Dogon drawing labeled Womb of All World Signs or Picture of Amma, which they employ as a diagram to explain the concept of the egg in a ball (see next page).

It is important to mention here the strong likelihood that the notion of an egg in a ball constitutes a common point of origin for two parallel symbolic themes of Dogon mythology—those of the formation of matter and of biological reproduction. Although in this study we have focused primarily on structure-of-matter symbolism, we should be aware of the second symbolic thread and how it may be represented by these same symbols. In this context, the Dogon drawing that Dieterlen describes as a picture of Amma is defined specifically as a womb; the circular figure at its center is described as an umbilicus.*

According to Dogon belief, Amma, who is a likely Dogon counterpart to the Egyptian god Amen, is made up of four attached clavicles (arcs) that together form an oval called "egg in a ball." This is the same term that is applied to the figure at the center of the nummo fish drawing.[6] Dogon tradition holds that this ball was "before all things," by which the Dogon priests specifically mean before space and time.

*Calame-Griaule also discusses the reproductive symbolism of Dogon symbols in *La Parole du Monde*. Specifically, she states that Amma creates a placenta and that the Dogon symbol of the fish can be compared to a fetus.[5]

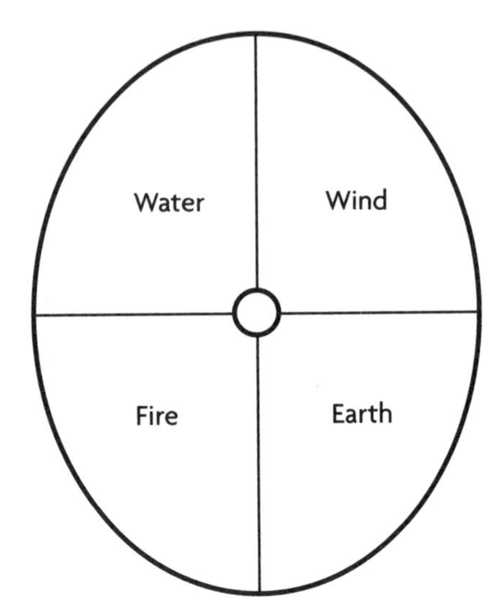

Dogon Womb of All World Signs, representation of Amma

According to Dogon myth, the four clavicles of Amma prefigure the four mythological elements of water, fire, wind, and earth, which we know based on our Structure of Matter table represent four component stages of matter. Likewise, the Dogon priests say that the pair of perpendicular lines—the *axis*—that demarcates these four clavicles will constitute the four cardinal directions of space.[7]

Based on ideographic readings, we can show that the Egyptian concept of a ball reflects symbolism that is similar to that of the Dogon. One Egyptian hieroglyphic word for ball is *pets-t,* a word whose two spellings depict both a ball and an egg in glyph form.[8] Ideographically, these two spellings relate to the concepts of space and the formation of matter. We perceive an idiom at work in these words; a verb preceded by the hand glyph 𓂝 seems to imply the past tense of the verb. For instance, the first reads 𓊪𓂝𓋴𓏏𓇳, which by our method would read "space gives bending to mass." However, the true scientific relationship according to Einstein is the reverse of this—it is mass that bends space-time. We can resolve this difficulty by postulating a convention of language that would allow us to read the word as "space is given bending (or is bent) by mass," defining the ball glyph. It is significant that the ball glyph is often used ideographically in the Egyptian language to represent the concept of seeds and in a larger incarnation as the circle glyph to depict an orbital path. The second spelling of the word *pets-t* requires the use of the same convention to sensibly satisfy its meaning. It reads □, meaning "space is bound (given binding) by the coiled thread," and defines the egg glyph ( ). In Dogon mythology, the concept of an egg refers both to the central oval in the womb of all world signs drawing and to the mythological counterpart of the Calabi-Yau space—the primordial egg that houses the seven vibrations of the coiled thread. The word *pets-t* derives from the phonetic root *pet* (□), which Budge describes as meaning "to open out, to spread out, to be wide, spacious, extended."[9] Ideographically, the word *pet* reads "space raised up (given raising up)."

The very same phonetic value "pet" also is used for the name of the Egyptian god Pteh or Ptah,[10] who according to Budge was "the great architect of heaven and earth . . . and the fashioner of the bodies of men."[11] In terms of the drawing of the womb of all world signs, Amma (or Amen) can be seen as being represented by the inner egg, and Ptah seems to be represented by the outer circle. Following our two parallel symbolic threads, the outer circle that is Pteh or Ptah corresponds both to the initial expansion of space at the time of the perception of matter and to the womb that encloses the egg during the process of biological reproduction. Ideographically, the glyphs that comprise the name Pteh support this same dual symbolism.

ARCHITECT OF HEAVEN, EARTH, AND FASHIONER OF THE BODIES OF MEN (PTEH)

The god Pteh	□	Space □, mass/womb, and DNA, followed by the god determinative. (See Budge, p. 254b)

Of particular importance to the Dogon in relation to the drawing of the womb of all world signs is the yu grain,[12] because its shape is said to resemble that of the union of Amma's clavicles (). A likely corresponding Egyptian word is *au,* which forms the basis of several Egyptian words relating to grains and the harvesting of grains (see the words *auhu*[13] and *aua/auai*[14]). We can associate these Egyptian words with the concept of the growth and bending of mass by way of the Egyptian word *au-t,*[15] which constitutes a defining word for the staff glyph (|) and which according to Budge means "stick with a curved end." Ideographically, the word *au-t* reads, meaning "act or force of the growth of mass," which can be seen as the bending of a string as it vibrates through the unseen dimensions of the Calabi-Yau space. This is the same concept that is symbolized in our Structure of Matter table by the staff glyph.

Guide Signs, Master Signs, and World Signs

Dieterlen's explanation of the egg in a ball drawing begins with the 266 seeds or signs that according to Dogon cosmology underlie all matter. The Dogon priests organize these signs conceptually into three groups: two guide signs, which lead to eight master signs, and then to 256 world signs. According to Dieterlen, these signs coexist in pairs; furthermore, the eight master signs, also in pairs, are associated with each of the four primordial elements: water, fire, wind, and earth. The signs in each pair symbolize opposing aspects of their associated element. For instance, the master signs associated with water represent moisture and dryness. Those associated with fire symbolize light and darkness (or alternately, fire and wood, which is what fire burns). Those associated with wind symbolize wind (air in motion) and air (still). Those corresponding to earth represent earth (with a small "e," meaning "mass") and empyreal sky (meaning "space").

The two guide signs (or literally, "eye-signs") represent the "springing forth of conception" and lead the way (literally, "show" or "make known" the way[16]) to the eight master signs. According to

Dieterlen, these two signs, which reside at the point of perception of the egg, *are* Amma. She tells us that before them, nothing else existed. We can see based on our familiarity with the Egyptian hieroglyphic language that these two original guide signs are also meant to represent opposing aspects of their associated concept, which in this case would be the concept of perception. Budge tells us that the word *am* means "to know"[17] and that the word *maa* means "to examine or perceive."[18] Even at these earliest stages of interpretation, we see evidence of two separate symbolic themes: metaphorically, knowledge (in the biblical sense) implies conception—the initiating act of biological reproduction—just as perception can be seen as the initiating stage of matter.

Dieterlen also tells us that the eight master signs are the eight ancestors of the Dogon. This statement, combined with the Dogon cosmological definition of the paired signs as representing opposing aspects of their associated elements, leads us directly to the eight paired gods and goddesses of the Egyptian Ennead: Geb and Nut, Shu and Tefnut, Osiris and Isis, and Set and Nephthys. These are the eight great ancestor gods of Egypt, whose traditional attributes seem closely linked with the four elements. For example, Geb traditionally is considered the Egyptian earth god, Nut is the sky goddess, and Shu is the god of air or wind.

The Dogon mythological keywords that form the foundation of the womb of all world signs—terms such as *guide signs, master signs,* and *world signs*—are well supported by words in the Egyptian language. They gravitate around the familiar hieroglyphic prefix *tch-*, a phonetic value that calls to mind the concepts of both the serpent  and the Word. If we begin with Budge's dictionary at page 899a, we see that the word for "guide" is *tchar,* a phonetic value that also forms the root of Egyptian words with meanings of both "sign" and "seed."[19] Another word with the same root sound, *tcharm,* means "plant"[20] and takes the form of a defining word for the three-stemmed plant glyph. (In the Freemason tradition, this same figure is used to symbolize the tree of life, whose roots were said to reach down into the watery abyss. The same tree of life concept is also closely linked to the notion of an axis, which is referred to as the *axis munde*. According to Budge, the Egyptian word for "tree of life" is *aam* or *aama*.[21] The word *aama* also qualifies as a defining word for the three-pronged plant glyph.)

Another Egyptian word *tchar* means "to spy or to scrutinize" and is a defining word for the eye glyph (𓁹), which is one of our interpreted symbols for the act of perception.[22] Yet another Egyptian word *tchar* means "to burn"—a likely reference to the mythological keyword *fire,* which we also take as a symbol for perception. We can see in these related Egyptian word entries confirmation of each of the key concepts expressed by the Dogon relating to the initial guide signs (or eye signs) of Amma. Clearly, the definitions of Egyptian words in the tchar/tcharm group are in agreement with the notion of guides, signs, eyes, seeds, and plants.

We find similar confirmation in the Egyptian language for Dogon references to master signs, stemming from the same root sound "tch." The word *tchaas* means "master."[23] Budge defines the word *tchaau* as meaning "a kind of seed or grain."[24] He also lists the words *tchaas*[25] and *tchab,*[26] which refer to a kind of plant and constitute alternate defining words for the same three-stemmed plant glyph. Based on these references and the appearance that each of these word groups that include the words *guide* and *master* define the same three-stemmed plant glyph, the suggestion is that three branches of the glyph may, in fact, symbolize the three sign groups defined in Dogon cosmology.

Dogon references relating to the third group of signs—the world signs—are also represented by Egyptian words beginning with the prefix *tch-*. The word *tchera* means "to be complete,"[27] and *tcher* means "all or whole"—specifically, according to Budge, in reference to all the gods.[28] Another word *tcher* means "a circuit" and constitutes a defining word for the ball or circle glyph ○.[29] Yet another word *tcher* means "to envelop."[30] It is also worth noting that the titles *Tcher* and *Tchera-t* mean "ancestress" and were assigned by the Egyptians to both Isis and Nephthys.[31]

As we have mentioned previously, there is yet another important Egyptian word reference with links to the Dogon concept of *amma*. Dieterlen states that the word *amma* means "to hold firmly, to embrace strongly and keep in the same place."[32] In the Dogon tradition, Amma is the god who holds the world firmly between his two hands; the Dogon say that when his name is spoken, Amma is entreated to continue to hold it. There is a corresponding Egyptian hieroglyphic word *am* or *amu* that means "to seize, to grasp."[33] This leads us to the Egyptian notion of *amen,* which means "to make firm, to establish."[34]

Waves Raise Up

As we have said, the process of the formation of mass is defined by the Dogon in terms of a four-stage creation process, which can be applied equally to the creative acts of Amma and those of mankind. These stages are called bummo, yala, tonu, and toy (or toymu). The Dogon define them using parallel metaphors, one of which is expressed in terms of the stages of building a structure or house.[35]

Cosmologically, from the moment that the perceived wave is caused to raise up, existence begins and the concept of space or place can be said to exist. This concept is reflected in Egyptian culture by the god Khet, who Budge describes as "the god of things that exist." Like the word *kheper,* the name Khet is formed from the phonetic root "ke," which we know in Dogon language connotes the idea of emergence from water. Another Egyptian word *khet* means "fire," which is a mythological symbol for perception[36] and is a defining word for the fire glyph. We also know that the Egyptian branch glyph is pronounced "khet"; it presents a pictogram for one of the opposing aspects of fire—wood, which is what fire burns. The Dogon myths tell us that the emergence of mass or matter is facilitated by the three categories of signs, a concept that may be conveyed ideographically by the glyphs in the Egyptian name Khet, which is also associated with the number three.

GOD OF THINGS THAT EXIST (KHET)

The god Khet	Source of mass or matter followed by the number 3 and the god determinative. (See Budge, p. 526a)

Hand in hand with the emergence of mass comes the appearance or existence of place or space. One of Budge's Egyptian word entries for place is pronounced "bu"[37] and is written with the single foot glyph, which is one of our ideographic symbols for the concept of place. Much as we have interpreted the Egyptian phrase *nu maa* to mean "waves perceived," so we also interpret the Egyptian phrase *bu maa*—the likely Egyptian counterpart to the Dogon *bummo*—as meaning "place perceived." The Dogon priests describe the mythological keyword *bummo* as referring to the "initiating stage of a creative act," a definition that is in agreement with the placement of the phrase *bu maa* within our Structure of Matter table. Based on that table, bu maa would represent the point at which the act of perception gives way to place or space. These interpretations are supported ideographically by the glyphs of Budge's Egyptian dictionary entry for the word *bu maa*.

PLACE OF TRUTH (BU MAA)

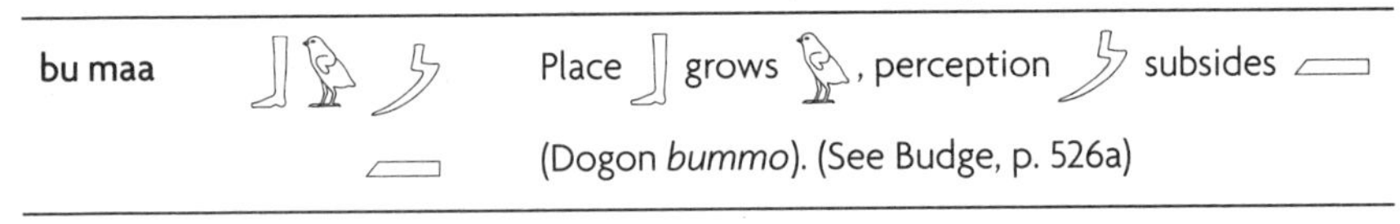

bu maa	Place grows, perception subsides (Dogon *bummo*). (See Budge, p. 526a)

This initiating stage in the formation of matter corresponds to the Egyptian birth goddess Ua (not to be confused with the Egyptian god Ua, who plays a role later in the formative processes of matter). Her name is expressed in terms of the glyph, which traditional Egyptologists interpret as a looped cord, symbolism that would be commensurate to the notion of growth of the coiled thread, a figure that traditional Egyptologists also interpret as a cord. Budge offers no explanation for the image presented by the glyph, but based on other growth-of-plant symbolism, it is possible that it depicts the raising up of a growing plant. In any case, ideographically we can see that the glyph may be meant to express the concept of growth based on the two explanatory glyphs included by Budge as a reference to its phonetic value[38], which we read ideographically as "growth comes to be."

THE BIRTH GODDESS (UA)

The goddess Ua		Growth/raising up comes to be, followed by the goddess determinative. (See Budge, p. 145a)

Delimitation Posts Form

Following the Dogon cosmological model, the next stage in the formation of matter is called *yala,* which means "mark, image, or sign." Dieterlen says that the sign of the house in Amma's body—prior to any manifestation—is referred to as the courtyard of the house.[39] The Dogon drawing that defines this courtyard takes a rounded form of the shape presented by the Egyptian courtyard glyph, which appears prominently in our Structure of Matter table. The development of the courtyard begins with the post of the house, which corresponds to the concept of delimitation posts depicted in our Structure of Matter table. The Dogon say that this post is in the domain of fire and wood[40]—a statement that makes complete sense in terms of our Structure of Matter table because, within the context of the cosmology, wood is considered to be the source of fire.

The Egyptian concept of delimitation posts is expressed by the word *ahau,* which we take as a counterpart to the Dogon word *yala*. It derives from the Egyptian phonetic root "aha," which is the same pronunciation that Budge assigns to the glyph, whose shape bears some resemblance to the body of the nummo fish. Budge defines the glyph as meaning to "stand up."[41] He also states that, in the Egyptian hieroglyphic language, the word *ah* means "to surround, to surround with a wall, to enclose."[42] The Egyptian word *ahi* refers to a "camp, courtyard" and qualifies as a defining word for the courtyard glyph—the same symbol we previously equated with the encircling wave that forms the divisions of Calabi-Yau space. Linguistically, this initiating stage of Calabi-Yau space corresponds to the related Egyptian word *auau,* meaning "jackal or dog," and so the jackal becomes the symbol of the Tuat or Calabi-Yau space. Prior to this point in the formation of matter, the massless wave can be thought of as being single, rather than dual, in nature, and so Amma's initial role in the formation of earth is seen as incestuous or masturbatory—symbolic of a sexual act involving only one partner. We have also seen that the massless wave as it draws up takes a peaked shape resembling an anthill, the very object with which Amma is said to have had incestuous intercourse in Dogon mythology. Likewise, we can think of the process of perception as disturbing the otherwise perfectly ordered state of the massless wave, and so the jackal comes to be associated with the concept of disorder. Ideographically, we can also see that the glyphs of the word *ahi* describe the notion of encirclement and growth.

A COURTYARD (AHI)

Courtyard		That which encircles growth, followed by the courtyard glyph. (See Budge, p. 74b)

Another Egyptian word *aha* is defined by Budge as meaning "wooden staff, prop, stick" and constitutes a defining word for the wood glyph .[43] Another defining word for the wood glyph is the Egyptian word *uatch,* meaning "stick, twig, pillar, support, column."[44] This word derives from the phonetic root "ua,"[45] which we previously have said means "one who becomes eight"—a reference to the first of eight encircling phases of the eight-stage Calabi-Yau space. Budge defines the same phonetic value "Ua" as the name of the One God, who he describes as "the number one of the gods."

Emanation

Next, according to the Dogon myths, the rising wave encircles and divides in two. Based on the various component elements within the nummo fish drawing, we see that this process takes the shape of the Egyptian teardrop glyph, the same shape that, according to the Dogon narrative, forms the two twin doors. These doors are used to represent the collarbones (clavicles) of the nummo fish and represent the point at which the rising wave becomes dual; this is the point at which mass in its simplest form appears. This third stage of the creative process described by the Dogon corresponds to the Dogon keyword *tonu* (from the Dogon word *tono,* meaning "to portray"). Following the metaphoric theme of building a house, the Dogon say that this conceptual stage is represented by an outline of pebbles, carefully placed at each of the corner points of the structure to denote where the future walls will be placed. The corresponding Egyptian word is *tennu,*[46] which Budge defines as meaning "border, boundary" and can be seen as a defining word for the courtyard glyph. Ideographically, the word *tennu* is written as follows:

DEFINITION OF THE COURTYARD GLYPH (TENNU)

Border, boundary		Gives waves existence, followed by the courtyard glyph and the numbers 1 and 3 determinatives. (See Budge, p. 881b)

In the Egyptian hieroglyphic language, the teardrop glyph is the sole glyph used to write the word *ta-t,* which means "emanation."[47] Another spelling of the same word *ta-t* is written, meaning "mass is." A third spelling of the same word can be seen as a defining word for the teardrop glyph and reads , followed by the teardrop glyph. It also means "mass is." Yet another Egyptian word, *tat,* means "room, chamber"[48] and constitutes a defining word for the courtyard glyph. It is written as follows:

DEFINITION OF THE COURTYARD GLYPH (TA-T)

Room, chamber		The drawing up and encirclement of mass , followed by the courtyard glyph . (See Budge, p. 821b)

The word *ta-t* comes from the phonetic root "ta." According to Budge, Ta was the primeval Earth-god, husband of the Sky-goddess.[49] A second definition describes Ta as "the Earth-god; the god of a Circle."[50] This name is written with the single glyph, which we take as an alternate symbol for mass that is expressed in terms of the encirclement initiated by the three categories of signs. Budge also defines the word *taa* as meaning "divine emanation."[51] The word *taa* is written or ideographically as "mass which has come to be," followed by the teardrop glyph and the god determinative. It is also worthy of note that the phrase Ta Tuat[52] qualifies as yet another defining word for the chamber glyph and is described by Budge as meaning "the land of the Other World." Likewise, Budge defines the word *Ta-t,* which is a synonym for *Tuat,* as "a very ancient name for the Other World."[53]

Another important Egyptian word for the concept of a door or gate is the word *re,* which is written . Another word *re* with the same spelling means "mouth," the pictogram that traditionally is assigned in common usage to the mouth glyph . By our previous interpretations, this glyph pictographically depicts the bending force of gravity, the force that Einstein believed causes orbits. If the passing of the perceived wave through the twin doors constitutes the first stage in the formation of mass, then it might also be said to represent the first appearance of gravity. On another level, the word *re* can be seen as the phonetic root of the

Egyptian word *Ra,* which refers to the sun, the day, and the Egyptian Sun god.[54] Another word, *ra,* means "work, act, action, to do."[55] The word *Ra,* which is also the name of the Egyptian sun god, means "the sun, the day" and is written as follows:

DEFINITION OF THE SUN GLYPH (RA)

The sun, the day		The bending/warping force, followed by the orbit and the number 1 determinative. (See Budge, p. 417b)

The Calabi-Yau Space Forms

The encircled wave that initiates the Calabi-Yau space chamber is represented in the Egyptian language as the god Sab, whose name in turn is represented by the single jackal or wolf glyph. Budge defines Sab as "the Wolf-god or Jackal-god."[56] The word *sab* means "wolf, jackal" and is a defining word for one of the jackal/dog/wolf/fox glyphs.[57] It is written, or ideographically as "binding of place," followed by the wolf/jackal glyph. Another word *sab* means "judge, chief, master." Sab-res is a title that refers to the god Anubis, whose name in the Egyptian language is Anp or Anpu.[58] Budge defines the word *anp* as meaning "to swathe, to wrap round,"[59] or ideographically, "that which weaves space," followed by the loop string intersection. Calame-Griaule says that the word *anu* in Dogon mythology represents "the seed which wraps around other seeds at the time of creation,"[60] a very apt description of the Calabi-Yau space.

Vibrations Begin in the Calabi-Yau Space

The relationship between the encircling of mass and the concept of vibration is a simple and direct one, expressed by the word *Tuat* and its phonetic root "ta." Budge defines the word *ta* as meaning "to tremble, to shake, to quake" and the word *ta-t* as meaning "trembling, quaking."[61] (If we follow along in the parallel theme of biological reproduction, we note that the word *ta* also refers to "the emission of a seed.") We have already discussed at length the many Tuat-related references to stars, rays of a star, and vibrations that correlate to cosmological descriptions of the Dogon. We know that "tua" is the phonetic value of the star glyph ☆[62]; we also know that the encircled star glyph ⍟, the same shape that corresponds to the Calabi-Yau space in Dogon cosmology, is used in the Egyptian language as a symbol for the Tuat and is explicitly defined by Budge as referring to "the Underworld, Tuat."[63] We have mentioned that the Egyptian deity Tua is the god of circumcision—a ritual cutting that calls to mind the tearing during the eighth stage of the Calabi-Yau space and that in Judaism occurs on the eighth day after the birth of a child.

Divisions of the Calabi-Yau Space

Each of the above references from the Egyptian language is consistent in the symbolism it assigns to the subset of glyphs that relate to the Tuat, and they support the notion suggested by our Structure of Matter table that we should view the Tuat as an Egyptian counterpart to the Calabi-Yau space. By that interpretation, the encircled star glyph ⍟ would represent what Dogon mythology clearly tells us it represents—the vibrations of a string within the Calabi-Yau space—and the chamber glyph ⊔ would represent one of the seven wrapped-up dimensions within the Calabi-Yau space. One Egyptian word for chamber, *urit,*[64] is based on the Egyptian phonetic roots "ur" and "urr." The first of these roots carries the meaning of "wind," which is a mythological keyword for vibration; the second means "to increase, to grow."[65]

Budge defines another set of Egyptian words that relate to the Tuat and are based on the root *art,* or *arit.* He defines *arit* as meaning a "judgment hall." This would seem like a very specific and appropriate type of chamber in which to house the judicial deliberations of the mythological wolf or jackal-god Sab, who we know is closely linked to the concept of the Tuat. Budge also defines the word *Arit* as a "division of the Tuat" and tells us that the Arits were seven in number,[66] the same as the number of wrapped-up dimensions defined in string theory for the Calabi-Yau space. The word *Arit* is based on the phonetic root "ari," which carries two definitions that are directly related to the formation of matter. The first definition is "wind," or as we interpret it, "vibration"; the second is described as "a kind of a fish," which calls to mind the nummo fish drawing of the Dogon. Another related Egyptian word is *ar-t*. It refers to the jawbone or the lower jaw and is a defining word for the Egyptian jaw glyph. This glyph takes the same general form as the sickle glyph but includes seven distinct teeth that are set into the jaw. The jaw glyph may constitute yet another glyph in an apparent progression from the adze, to the sickle, to the jaw glyph, each symbolizing by its shape progressive stages in the formation of mass.

Budge tells us that each of the seven Arits was in charge of (Budge specifically says "in charge of," p. 130A) a doorkeeper, a watcher, and a herald. The concept of a door in regard to the formation of mass has already been defined during our discussion of the nummo fish drawing, in which the twin doors appeared to represent the gateways to the twin Calabi-Yau spaces. So it is reasonable to presume that a door could also represent the gateways between the seven encircling chambers within the Calabi-Yau space, the wrapped-up dimensions through which a string must bend on its journey to create mass. One Egyptian word for doorkeeper is *arit-aa*. It is written as follows:

DEFINITION OF A DOORKEEPER (ARIT-AA)

Doorkeeper		That which bends mass, followed by the door glyph. (See Budge, p. 70a)

The Egyptian for the word *watcher* is based on the root word *res,* which means "to watch" and seems closely related to the concept of perception, which is the initiating act by which mass is first created. The word *res* is written and consists of three symbols that by their very shape seem to convey the notion of bending.[67] The word *resi* means "wind"—again by our interpretation, "vibration"—and is a defining word for the wind glyph ; the word *resres* means "to build."[68] So if the doorkeepers guard the passages between each of the wrapped-up dimensions of the Calabi-Yau space, then the watchers, who are harbingers of perception, could easily be interpreted as overseeing the vibration and bending of the string by which mass is formed within these dimensions.

One of the Egyptian words for herald is *uhemu,* meaning "teller, proclaimer, or herald."[69] This word is based on the root word *uhem,*[70] which means "to repeat." Budge interprets another word *uhem* as meaning "to renew, to repeat an act."[71] Another word, *uhem-t,*[72] means both "what is repeated, something that is renewed," and "a revolution of a star." However, we know that the Dogon mythological concept of a star corresponds to the seven windings or revolutions of a string as it passes through its repeated vibrations in the Calabi-Yau space. So we can interpret the Egyptian concept of herald as representing the repeated windings of the mythological space as the vibrating string passes through the seven Arits of the Tuat.

The Coiled Cosmic String Exists

We have also discussed at length the coiled thread and its relationship to the concept of vibrations or trembling. Egyptologists interpret the coiled thread glyph as representing a cord. We gain insight into the possible deeper meanings of the symbol when we read Egyptian words for "cord" ideographically. One such Egyptian word, *un-t,* is a defining word for the coiled thread glyph and is written as follows:

DEFINITION OF THE COILED THREAD GLYPH (UN-T)

Rope, cord		Vibration creates mass , followed by the coiled thread glyph . (See Budge, p. 167a)

We know that the concept of vibration is expressed by the Egyptian phonetic value "un" and that one Egyptian word meaning "to tremble" is *unun.*[73] We can relate this notion of trembling or vibration with the concept of existence through the name of the Egyptian god Un, who is defined by Budge as "the god of existence."[74] Likewise, the concept "to be, to exist, to become" is expressed by ideographic images of vibration, waves, and the coiled thread in the Egyptian word *un* or *unn.*[75]

Mass Is Complete

The fourth major stage of the creative process, according to Dogon cosmology, is called *toy,* or *toymu,* which means "complete." It is a likely counterpart to the Egyptian word *temau,* which Budge defines as meaning "all, complete."[76] A corresponding Egyptian god would be Tem, or Temu, who Budge describes as "the creator of heaven and earth."[77] He assigns this same phonetic value, "tem," to the sledge glyph [78] and includes in his definition the referential glyphs , which we interpret ideographically to mean "mass known."

The Calabi-Yau Space Tears

According to Dogon cosmology, after the seven vibrations of the cord within the Calabi-Yau space, which the Dogon describe as rays of a star, have been fully encircled, the last of these rays grows long enough to pierce the wall of the "egg." Corresponding concepts are represented in the Egyptian hieroglyphic language by words that are pronounced "set." According to Budge, the word *set,* which means "encircled,"[79] can also mean both "to cut, to pierce" and "to breach a wall."[80] Similar concepts are conveyed ideographically by the name of the Egyptian god Set, who Budge defines as "the god of evil," a reputation perhaps assigned based on his symbolic role in the demise of the completed Calabi-Yau space.

DEFINITION OF THE STABBING/PIERCING GLYPH (SET)

The god Set		The bending of mass complete/whole, followed by the stabbing/piercing glyph. (See Budge, p. 706b)

Another Egyptian word *Set* refers to "the star of set"[81] and provides us with a counterpart to the Dogon image of bending mass as represented by the rays of a star.

DEFINITION OF THE STAR GLYPH (SET)

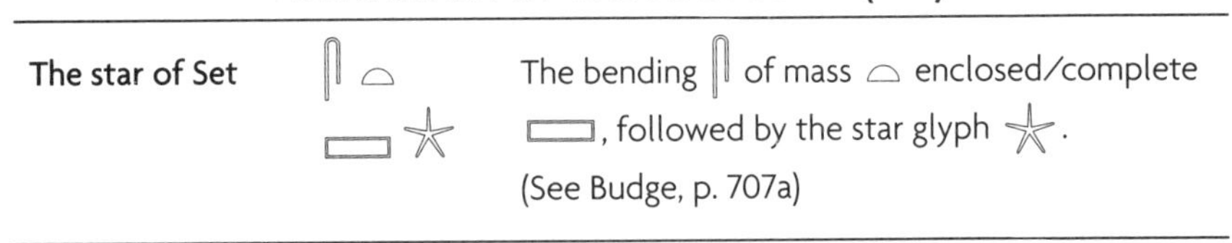

The star of Set		The bending of mass enclosed/complete, followed by the star glyph. (See Budge, p. 707a)

The Word Is Spoken

Dieterlen tells us that the eighth stage of the cosmological egg—the piercing of the wall—constitutes the final articulation of the Word in Dogon cosmology.[82] This event represents the closing stage in each of the extended metaphors used to describe the growth of vibrations within the egg: stages of the building of a house, descriptions of the growth of a star, and descriptions of the formation of the Word. Each of these three concurrent themes comes together in the Egyptian language in forms of the word *Tchet,* which according to Budge represents "the Divine Word, speech deified."[83]

From a parallel symbolic perspective, the word *tchet* can be seen as representing a culminating stage in the process of biological reproduction. For Budge, another Egyptian word, *tche-t,* means "body, person, bodily form"[84] and in fact is defined by him to mean "Divine Body"—the biological correlate to the cosmological "Divine Word."

The mythological notion of the Word is symbolized in the Egyptian language by the serpent glyph , to which Budge assigns the phonetic value of "tch."[85] From this perspective, Budge tells us that the word *tche-t* can also mean "papyrus," which is the medium of the written word.[86] Looking further, we see that the serpent glyph is defined in the word *atchet,* which means "to make a reply, to speak."[87] This word is written as follows:

DEFINITION OF THE SERPENT GLYPH (ATCHET)

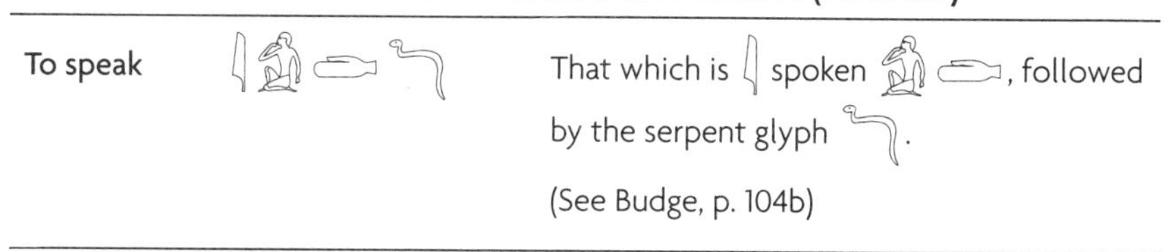

To speak		That which is spoken , followed by the serpent glyph . (See Budge, p. 104b)

There is another Egyptian word pronounced "tche-t" that Budge defines as meaning "place, house, abode" and that qualifies as yet another defining word for the courtyard glyph . Appropriately, it is expressed ideographically in terms of *mass* .[88] A related word *tchet-t* means "star, time of the culmination of a star"[89]; the word *tches* means "to cut, to divide."[90]

The Egyptian concept of the Word centers on the highly significant root word *per,* which means "what comes forth from the mouth, i.e., word, speech."[91] Cosmologically speaking, the Word is formed at the point where the bendings and windings of the Calabi-Yau space have reached their conclusion. The word *per* is written as follows:

WHAT COMES FORTH FROM THE MOUTH (PER)

Word, speech		The structure of bending spoken . (See Budge, p. 240b)

From an Egyptian perspective, we can relate the concept of the spoken word to the Tuat—the counterpart of the Calabi-Yau space—by way of the word *per-tuat*.[92] In its shortest form, this word is written (ideographically, "chamber of vibrations") and refers to a "chamber of the Other World." A longer spelling of the same word can be seen as a somewhat self-redundant defining word for the chamber glyph and is written as follows:

DEFINITION OF THE CHAMBER GLYPH (PER-TUAT)

Chamber in the Other World		Chamber gives growth to vibrations of mass , followed by the chamber glyph . (See Budge, p. 240a)

The root *per* is significant in this context, because it forms the basis of the Egyptian phrase *per em hru,* which means "coming forth by day" or "coming forth into day,"[93] which is the title of an important Egyptian text commonly referred to as *The Egyptian Book of the Dead,* also known as *The Book of Coming Forth by Day.* The word also plays a similar role in the phrase *per her ta,*[94] which means "to appear on earth, i.e., to be born." Again, we are made aware of two parallel lines of symbolism defined by the same mythological keywords and symbols: one pertaining to astrophysics and the other to biological reproduction. The suggestion is that an ideographic reading of *The Egyptian Book of the Dead*—known to the Egyptians as *The Book of Coming Forth by Day*—may yield details about the processes of the Tuat in its role as counterpart to the Calabi-Yau space.

Matter Is Woven

The concept of the weaving of matter is not one that constitutes its own discrete stage, even though for the sake of simplicity it is depicted as such in our Structure of Matter table. In fact, the symbol of the looped or self-intersecting string is one that we can see, based on discussions in this chapter, and may actually be associated with the encirclement of the wave as it forms the Calabi-Yau space. Likewise, the symbol of two intersecting strings seems to be associated with the final stage of the Calabi-Yau space, during which the "wall of the egg" is pierced. The more complex string intersection glyph comes into play during the formation of membranes, which are symbolized by the serpent glyph . This entire process of the weaving of threads is associated with the goddess Neith, who was the great mother goddess and the weaver of matter, and it actually spans each of the preceding stages of matter from just after the point of perception to the point of the spoken word. From that perspective, these other stages of the formation of mass are under her direction, as are the later component stages as well. Therefore, each of the Egyptian gods and goddesses after Amen, Ptah, and the eight Ennead ancestor deities can be seen as true *neteru,*[95] that is, gods and goddesses who descend from the goddess Neith. However, as we will see later, Dogon cosmology provides us with even more specific and compelling reasons to see her in the role of the primordial Mother.

The Egyptian goddess Neith corresponds to the spider Dada in Dogon cosmology, whose name, we are told, means "mother" in the Dogon language. Both characters are seen to represent the Great Mother who spun and wove the threads of matter. Although in this specific instance there is an obvious difference in name between the characters of the two mythologies, Dada and Neith (and her later Greek counterpart Athena) both do carry spider symbolism, which would thereby seem to have been an original part of the Egyptian cosmological system. The appearance of a spider at the lowest levels of the formation of matter is entirely in keeping with the metaphoric cosmological mind-set that runs from insect to fish to four-legged animal to flying bird.

Quarks Are Formed

The concept of quarks seems to be represented in the Egyptian hieroglyphic language by the double-handed glyph , which Budge pronounces "ka." He tentatively defines the glyph as meaning "the double" or "person."[96] This would seem like an appropriate identification for a component of matter such as quarks, which occur naturally in pairs. Budge defines the Egyptian god Ka as the "god of letters and learning."[97] The name of Ka is written with the bull glyph  , a figure whose horns re-create the U shape of the glyph. Budge includes two glyphs as a reference to the phonetic value of the bull glyph  , which we read ideographically as "collected mass comes to be."

Quarks are subcomponents within the structure of matter that are bound into larger bodies such as protons and neutrons by the weak nuclear force, which is symbolized in our Structure of Matter table by the figure of a bowing man  . This figure correlates to a stage of Dogon cosmology that is defined by the phrase "that bows its head." Appropriately, there is another of Budge's Egyptian word entries pronounced "ka" that means "to bow."[98]

Protons, Neutrons, and Electrons Are Formed

Protons, neutrons, and electrons are the components of matter that correlate to the Dogon cosmological concept of sene seeds. They are symbolized in the Egyptian language by the word *senu,*[99] meaning "pot, vase, vessel, jar," which describes our interpreted symbol for a particle. Metaphorically, if creation results from waves or water, then the symbols for collected mass  and particles would be appropriately expressed by images of vases and pots. The same imagery is explicitly reinforced in the surface story line of the Dogon, in which the planets are compared to pellets of clay and the sun—in essence, a solar-sized particle—is compared to a clay pot. According to Budge, the Egyptian word *sen* means "clay."[100] The word *senu* is written as follows:

DEFINITION OF A PARTICLE (SUNU)

Pot, vase		Bound particle. (See Budge, p. 605b)

As we have mentioned, protons and neutrons consist of different combinations of three quarks bound together by the weak nuclear force, which we said equates with the Dogon phrase "that bows its head." Another Egyptian word *sen* is defined by Budge as meaning "to bow, to pay homage, to entreat."[101] It is written as follows:

DEFINITION OF THE BOWING MAN GLYPH (SEN)

To bow or pay homage		Binding force, followed by the bowing man glyph. (See Budge, p. 603a)

Budge defines yet another Egyptian word entry *sen* as meaning "to copy, to make a likeness or transcript of anything"[102] and a related word, *sennu,* on the same page as meaning "likeness, image, copy." The same pronunciation "sen" is applied to the oval glyph ,[103] to which Budge assigns the meanings of "go, pass, like, similar." The oval glyph is an unusual character among the body of Egyptian hieroglyphs because it carries two distinct phonetic values—"sen" and "au." We have demonstrated how the keyword *sen* relates to the concepts of protons and neutrons. It is significant that the second phonetic value of "au," which is the root sound of the word *aun,* is the very one we assigned to the concept of the electron in *The Science of the Dogon.*[104] The Dogon describe the path of their mythological sene seed, which is the counterpart of the electron, as a nest. The Egyptian word *au* means "nest, home" (see Budge, p. 32a), as does the word *aunnu.*[105] The Egyptian word *aun* means "to open, to make to be open"; its spelling includes the nest glyph —the very image of a typical electron orbital pattern.

CONTEXTUAL DEFINITION OF THE NEST GLYPH (AUN)

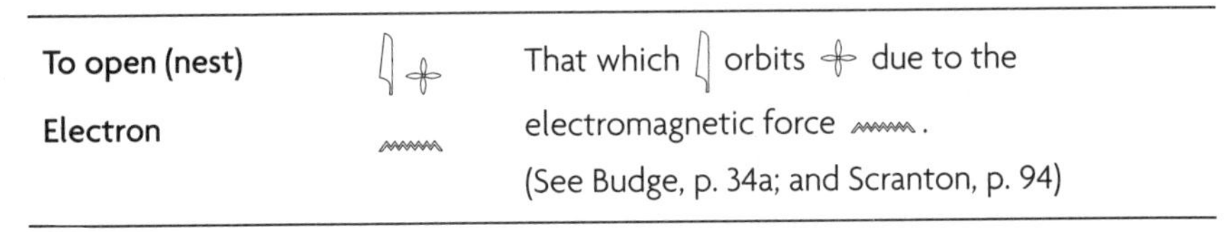

To open (nest)		That which orbits due to the
Electron		electromagnetic force .
		(See Budge, p. 34a; and Scranton, p. 94)

It is altogether fitting that another Egyptian word meaning "to open" is also pronounced "sen."[106] From each of the above relationships, we see that the Dogon mythological concept of sene seeds, which correlate to protons, neutrons, and electrons, is affirmed in several different ways by Egyptian language references that cause us to associate the phonetic values "sen" and "au," both of which are symbolized by the single Egyptian oval glyph . It is within this context that the Egyptian word *Senu* can be seen as referring to "a company of gods who fed Ra."[107] Ra, of course, represents our interpreted symbol for the bending force of gravity, which is "fed" by particles of mass.

The Atom

From many different perspectives, the Dogon symbol of the po—the mythological counterpart to the atom—represents the culminating stage of symbolism. The Dogon describe the po as "the image of the origin of matter."[108] They say that all things begin with the po and form themselves by the continuous addition of like elements. In the Egyptian tradition, Pau was the primeval god of existence. Budge says that his name may mean "he who is, he who exists, the self-existent."[109] The name Pau comes from the phonetic root "pa," which means "to be, to exist."[110] As we have previously mentioned, the word *pau-t* refers to "the matter or material of which anything is made"[111] and defines the hemisphere glyph ⌓, which we associate with the concept of mass.

Dieterlen tells us that the word *po* comes from the same root as *polo,* meaning "beginning." Budge tentatively defines another Egyptian word *po* as meaning "primeval time." He also refers to the Egyptian god Pauti as "the primeval god, the god who created himself and all that is." [112]

In regard to the second symbolic theme of the Dogon myths—that of biological reproduction—it is significant that another Egyptian word *pa* also refers to an "ancestor"[113] and that the word *papa* means "to bring forth, to bear, to give birth to."[114] From these references, we can see that, just as the concept of po or pau represents the final component stage in mythological structure of matter symbolism, it also constitutes the final stage of biological symbolism—the birth of an infant.

In the Egyptian hieroglyphic language, the phonetic value "p" is carried by two glyphs—the square glyph □, which by our interpretation represents space, and the flying goose glyph, our symbol for the atom. By any interpretation, the atom represents one of the fundamental delimiters of "space," and the image of a flying goose constitutes the final stage in growth-of-bird symbolism in which a bird begins as an egg, emerges as a chick, appears as a standing goose, and ultimately is set free, symbolically, as a flying goose. This symbolism is underscored by another Egyptian word *pa,* which according to Budge means "to fly" (see Budge, p. 230a). Likewise, in the Egyptian hieroglyphic language, *pait* are "feathered fowl, birds, water fowl" (see Budge, p. 230a). So the final stage of matter that emerged from water culminates with the symbol of a goose that literally takes flight above the water. The same symbol of a flying bird serves as the last of four stages of symbolism that began with an insect, progressed to a fish, transformed into a four-legged canid, and emerges as a flying bird.

Last but not least, Budge tells us that the Egyptian word *papait* refers to "a kind of grain or seed," which is the end product of the growth-of-plant symbolism shepherded by the sickle glyph. This symbolic theme began with a seed that rose up as a sprouting plant and became a grown plant, and it finally reaches culmination in the symbol of the stepped pyramid, the likely Egyptian counterpart to the Dogon granary. And so the symbol of the po or pau as it is expressed in Egyptian hieroglyphs, like each of the other words and symbols of the Egyptian language, affirms and supports each of the key symbolic themes of Dogon cosmology.

THIRTEEN

CONCLUSION

The preceding chapters present a large volume of information, some of it old, some of it quite ancient, some new, and much of it subject to new interpretation. The interpretation is different primarily because the point of entry to the Egyptian language is different. In this study our frame of reference does not center on the Rosetta stone and its comparative language texts, but rather on the myths and symbols of the Dogon—a modern-day African tribe whose culture exhibits marked similarities to that of ancient Egypt. This difference in approach carries with it a subtle but important shift in orientation, one that draws us away from the word-based equivalencies of Jean-Francois Champollion, the man who first deciphered hieroglyphics, and toward what are essentially symbol-based comparisons based on Dogon drawings and Egyptian glyphs.

In the traditional view of language, letters are the carriers of phonetic value and words are the carriers of concepts. And so it is entirely understandable and appropriate that Champollion, in his remarkable efforts to translate the Egyptian hieroglyphs, would assign meaning to sets or blocks of Egyptian glyphs—those groupings that most closely corresponded to known words in the comparative texts. However, based on interpretations of this study, it appears more likely that within the Egyptian hieroglyphic language phonetic values may be carriers of root concepts, including such fundamental notions as the act of perception or that of making, doing, or growing and concepts such as space and mass. Each of these phonetic values, when assigned to a glyph, would endow the glyph with its associated meaning. We can be confident that the more fundamental relationship between root concept and phonetic value preceded the creation of the glyph itself because in many cases we have shown that we can predict a glyph's ideographic symbolism based solely on its phonetic value.

Moreover, the root concepts that we associate with some Egyptian phonetic values appear to be additive. We have shown that we can combine a phonetic value such as "n," meaning "wave", with the phonetic value "u," meaning "to grow", and produce "nu," meaning "wave grows", a phonetic value that is represented by the adze glyph and that for us symbolizes the growth of mass from primordial waves. Furthermore, each glyph takes on added dimensions of meaning based on the image it depicts. For example, a picture of a rabbit or hare brings with it a sense of twitching or trembling, just as the figure of a falling man conveys a clear notion of the effect of gravity. From this perspective, if each Egyptian glyph corresponds to a word, phrase, or concept, then each hieroglyphic word, in essence, becomes an ideographic sentence. One key aspect of the Egyptian language that we have demonstrated again and again in various examples throughout this study is that these hieroglyphic word-sentences consistently define their own meanings.

The idea that phonetic value should be a determiner of meaning is one that is reflective of nature and science according to the precepts of string theory, which proposes that what we perceive as the different particles of matter actually derive their attributes from the different vibratory patterns of underlying strings or threads—in essence, in the view of string theory it is *sound* that determines *substance*. And so in the Egyptian hieroglyphic language the process of the formation of mass is equated with the speaking of a word. However, there is a grand

and elegant symmetry to the broader scheme of the language, which, in reflection, appears to use the structure of matter itself as the foundation for written language.

In my examples, I have shown that some categories of Egyptian hieroglyphic words and glyphs perform special functions. The first of these are what we have called defining words, which establish the ideographic meaning of a specific glyph. These words are easily recognized: their common-usage meanings describe the object or action depicted by the glyph, and the defined glyph appears as the final nondeterminative glyph of the word. Often, in pure examples, these words will carry the same pronunciation as the phonetic value of the glyph they define.

A defining word may express its meaning in several ways: 1) purely by way of context (like the word for high or rising water), creating what we might call a contextual definition, 2) through an ideographic sentence (like the word for time, which reads "that which is bent by mass"), or 3) through a list of examples (like the word *remen,* which means "to bear, to carry, to support, to hold up"), forming what I call an enumerating definition. An enumerating definition may consist of an ideographic sentence whose meaning defines a group of related glyphs, or it may consist of a list of related glyphs used to define another single glyph. An example of a possible fourth category of defining word is found in the word *techera-t,* meaning "kite."[1] This word appears to define the kite glyph  , which I interpret as meaning "to become" or "to come to be." Ideographically, the word reads "as bending/warping is to mass." Of course, bending and warping is the process by which mass comes to be, so a definition is conveyed by way of direct metaphor. I refer to this as a metaphoric definition. And so in the Egyptian hieroglyphic language, just as an ideographic sentence can define a word, so a word may also define an individual glyph.

Glyph meanings can be explicated further by examining an apparent relationship between the phonetic value of a glyph and an Egyptian deity whose name is pronounced like the glyph. The traditional role of the deity in Egyptian mythology appears to provide us with explanatory detail about the meaning of the glyph. The existence of a similarly pronounced deity name marks, or flags, the glyph as an important component (or component stage) in one or more of the creation themes of the mythology.

According to the entries in Budge's dictionary, some Egyptian words can be written using a single glyph. In these cases, Budge's definition for the word can often be taken as an ideographic definition of the glyph. One cautionary note is that, as we have seen, any given glyph can carry more than one related ideographic meaning, and so the glyph can appear as the object of several different defining words. This same multiplicity of meaning is acknowledged in more mainstream views of the Egyptian hieroglyphic language. As a case in point, the sun glyph ☉ traditionally is seen as representing the sun, is used to write the word *day,* and is taken to imply the notion of a unit of time. However, when we examine these concepts from an ideographic perspective, we begin to see that these relationships may not be arbitrary ones. By our interpretation, the sun glyph depicts an orbit—more specifically, Earth's orbit around the sun. In that context, the dot at the center of the glyph would represent the sun. Likewise, the period of Earth's orbit around the sun *does* constitute a fundamental unit of time (a year), and the notion of a day *is* defined by a relationship between Earth and the sun (the period of rotation of Earth in relation to the sun).

We can see that the assignment of symbols to concepts in these mythological systems could not have been the product of casual adoption by an unknowing populace. This observation comes sharply into focus when we examine the use of the scarab or dung beetle, known as the kheper, as a symbol for the creative source. What we see are several direct and compelling reasons for the selection of this specific symbol to represent this specific concept. First, we know that the Dogon use the word *ke* to refer to the dung beetle (in specific) and to all water beetles (in general) and therefore that the Dogon consider the dung beetle to be a creature of the water, which is the primordial element. Second, we recall that our entry point to Dogon cosmology was the central figure of the nummo fish drawing—the womb of all world signs—which is described as an egg in a ball. Of all the creatures of the earth, the dung beetle is one that appears to actually lay its egg in a ball—albeit a ball of dung. Third, the oval back of the scarab is recreated in the shape of the nummo fish drawing. The markings on the back of the scarab take the rough form of a *Y* or a *T*, which is replicated in the shape of the clavicles or twin doors within the nummo fish drawing that defines a wave as it rises up. Also, the scarab's head resembles the hemisphere glyph ⌓, which is our ideographic symbol for mass. Fourth, the Egyptian hieroglyphic language appears to use progressive images from the animal kingdom as yet another guiding metaphor for the formation of matter, in a manner similar to mythological references to water, fire, wind, and earth. This progression ranges from insects to fish to four-legged animals to flying birds. And so the dung beetle—an insect—can be said to belong properly at the earliest stage of the formation of matter. Fifth, based on our interpretations of the Egyptian language, the phonetic value "kh" means "source," an implied "e" carries the implied meaning of "and" or "of," "p" means "space," and "r" implies "time" or "gravity." So phonetically (shall we say "ideophonically"), the word *kheper* would imply the concept "source of space and time," and its longer form, the name of the god Khepre, could be interpreted as "source of space, time, and gravity."

Ideographic root words, which in our view preceded the appearance of written language and so seem to have formed their relationships with other words based first on pronunciation, do not follow traditional glyph-based rules and may not always depend on common sets of written glyphs for their interconnections. Nor do subtleties of vowel-sound vocalization (long "a," short "a," dotted "a," and so on) appear to have significant bearing on these phonetically based relationships. For instance, the mythological link is clear between the Dogon concept of Amma—the god who is entreated to continue holding the world between his hands—and the concept of clay—the material that Amma is said to have "thrown out" to create the planets and stars. For Budge, the concept of "entreating" is expressed by the word *amma* (see Budge p. 50b) and is written with the reed leaf glyph, whereas the word *ama,* meaning "clay" (see Budge, p. 122a), begins with the bent arm glyph.

An ideographic approach to the Egyptian hieroglyphic language simplifies, somewhat, the concept of the determinative because it removes from that category of language the poorly understood set of trailing glyphs that by our interpretation are seen as defined glyphs. One traditional explanation for the appearance of these glyphs at the ends of words is that they are "drawn into the word" as determinatives for emphasis, and so the word for *adze* would exhibit the adze glyph as its final character to reinforce the meaning of the word. With the defined glyphs removed from consideration as determinatives, we are left with two major classes of true determinatives—those in the category of the god or goddess glyphs and those that constitute numbers. We can infer the meaning of the god or goddess glyph by its use in relation to components in our Structure of Matter table. The entries in the table establish a clear pattern; the Egyptian words that carry the god or goddess determinative are those that define the glyph's component role within the larger structure.

Number determinatives are somewhat more subtle in their function, primarily because we are not always certain to which aspect of the word's meaning the number refers. By the traditional view, a trailing number three implies a word in the plural. However, in the word *sen,* meaning "they, them, their" (see Budge, p. 603a), which we take as a cosmological keyword for protons and neutrons, the number three determinative is a likely ideographic reference to the three quarks required to form a proton or neutron. In the word *ua,* meaning "only one" (see Budge, p. 153a), we interpret the trailing number one as a reference to the first of eight stages of vibration. In general, the determinative number serves in much the same way as the number notation on a modern-day street sign, which indicates what series of address numbers will appear in the next block. Number determinatives appear to provide an orienting reference, one whose effect may be to reassure us about the correctness of our interpretation of the word.

Based on examples within this study, we realize that in ideographic terms, god glyphs do not indicate gods or goddesses in the traditional religious sense, nor could the Egyptian gods and goddesses themselves have been intended to represent actual deities. In *The Pale Fox,* Marcel Griaule and Germaine Dieterlen tell us that the eight master signs within the womb of all world signs drawing represent the eight Dogon ancestors who correlate to the eight Egyptian Ennead gods and goddesses. Each is the personification of a component stage in a larger scientific structure, and their mythological attributes are careful descriptions of the corresponding stage of creation.

However, we do not have to rely on the Dogon priests alone for this interpretation. Based on our own interpretive method, we would expect the Egyptian words for god to offer an ideographic explanation of the concept of a god. The most direct of these is the Egyptian word *neter,* which can be written using the single flag glyph.[2] To flag, of course, is what the god and goddess glyph determinatives do for words that describe component structures of matter. An expanded form of the same word *nether,*[3] meaning "god," explains the concept ideographically. The meaning of the word becomes more understandable if we consider Budge's word entry for *th,*[4] in which the glyph is equated with the hemisphere glyph , our symbol for mass, and the word *neth.*[5] The entry states that the glyphs are considered to be an equivalent for the glyphs . The word *nether* is written:

DEFINITION OF A GOD (NETHER)

God		Flags how waves and mass bend and grow from seeds. (See Budge, p. 408b)

Ideographically, the word *nether* defines the concept that underlies the use of the flag glyph and god or goddess determinatives to mark component stages in the growth of waves from seeds to mass. However, what is not addressed by this word is the larger notion of god as a religious concept. For that interpretation, we must look to another Egyptian word, *khem,* which we interpret ideophonically to mean "source of completion," and which Budge defines as "God; he whose name is unknown, he who is not known."[6] Ideographically, the glyphs of the word *khem* read as follows:

DEFINITION OF GOD (KHEM)

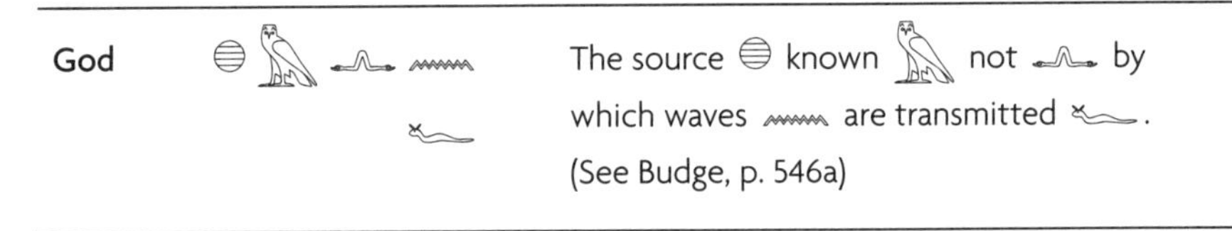

God		The source known not by which waves are transmitted. (See Budge, p. 546a)

There are several significant aspects to this reading of the word *khem.* First, we can see that it expresses the notion of god, not only in distinctly scientific terms, but also in terms that are entirely consonant with traditional religious views. God is defined as the unknown source of creation or, from a scientific perspective, the unknown primordial source of massless waves. However, what is perhaps even more significant from an interpretive viewpoint is all that the word does not say. It does not define god as an ancestor/teacher, as the architect of ancient language, or even as the builder of the pyramid or the Sphinx, nor does it define god as one of the traditional mythological gods of ancient Egypt, such as Amen, Ra, or Atum, or even as the great Mother goddess Neith.

Yet, within the larger Dogon and Egyptian cosmological structures presented here, we see the likely roots of a host of familiar elements found in many different modern religious traditions of the world.

The word *khem* held such significance to the Egyptians themselves that ancient Egypt was referred to as the Land of Khem. Likewise, according to Budge the word *khemen* means "eighth," a likely reference to the eighth stage of completion of the Calabi-Yau space.[7] Appropriately, the eight elemental deities were referred to as Khemenu.[8] Phonetically and ideographically, we can see that the word *khem* is closely associated with the concept of completion. Another Egyptian word *khem* means "to feign ignorance, to play the fool."[9] As previously noted, these two notions together—those of completion and of feigning ignorance or remaining silent—are the two actions incumbent upon an initiate in the most ancient esoteric traditions and constitute two explicit meanings of the word "Dogon," which in the Dogon language means "to complete the words" and "to remain silent." This same concept of silence is expressed by the Dogon word *keme,*[10] which is a likely counterpart to the Egyptian *khem.*

Also defined within the Dogon and Egyptian cosmological structure are other key concepts that are shared by modern religions. For example, we can see based on the dual themes of Dogon mythology that the concept of Divine Emanation refers to two events: the separation of a wave after the act of perception and the division of an egg after the act of conception. Likewise, we can interpret an ideographic concept of manifestation based on the Egyptian word *per-t,* which according to Budge means "exit, issue, what comes forth, manifestation, outbreak of fire, offspring."[11] We know that on one level this word refers to transmission of mass at the time of the completion of the Calabi-Yau space because of the ideographic form of the root word *per* 𓂻𓂻𓂋, meaning "to go out, to go forth."[12] However, it becomes apparent based on Budge's added meaning of "offspring" that it may also refer to the notion of birth—the coming forth of a baby within the context of the mythological theme of biological reproduction. The word *per-t* is written as follows:

DEFINITION OF MANIFESTATION (PER-T)

Manifestation (To go out, To go forth)	[hieroglyphs]	The coming forth [glyph] of gravity [glyph], mass [glyph], and the coiled thread [glyph]. (See Budge, p. 240b)

The completion of the coiled thread, which signals the appearance of mass and gravity, is an event that occurs at the completion of the Calabi-Yau space, which for the Dogon is considered to be the Second World. It corresponds to the Egyptian Other World or Underworld—the Tuat. Traditionally, the Tuat is intimately associated with Egyptian notions of life, death, and resurrection and, in the view of traditional Egyptologists, is the subject of the previously mentioned *The Egyptian Book of the Dead,* also known as *The Book of Coming Forth by Day.* It is worth reemphasizing here that based on an ideographic reading of the concept of going or coming forth, the suggestion is that this text may actually have to do with the formation of mass within the Calabi-Yau space.

The many similarities that exist between Dogon and Egyptian cultures, rituals, myths, symbols, and languages suggest that at some point in the distant past, there may have been a close relationship between the two groups. There are also signs that this relationship may date from early in the history of ancient Egypt. For example, although many of the Dogon cosmological drawings take the same shapes as Egyptian glyphs, the Dogon priests do not employ these shapes in the context of an actual written language. Nor is it really thinkable that the Dogon—a culture that places a high value on the purity of language—could have simply lost their cultural memory of a written language over time. We know that the hieroglyphs made their appearance early in the history of ancient Egypt. These simple facts alone would suggest that any Dogon interaction with ancient Egypt is likely to have occurred prior to the formalization of written language in that region. The same suggestion would be consistent with what we observe to be true about other Dogon symbols. For instance, in Dogon cosmology, the eight master signs have acquired status as representing paired ancestors and symbolism identifying them as opposing aspects of the primordial elements water, fire, wind, and earth, but they have not taken on the names and attributes of the corresponding Egyptian Ennead deities. Likewise, the Dogon granary is defined in terms that are reminiscent of the pyramid, but it has not taken on an actual identity as a pyramid.

This tentative placement of Dogon roots at the earliest days of Egyptian culture is strongly supported by a study written by North African anthropologist and ethnologist Helene Hagan titled *The Shining Ones: An Etymological Essay on the Amazigh Roots of Egyptian Civilization.* As we have previously noted, the Amazigh were the hunter groups (ancestors of the Berbers) that resided in the Nile Delta region prior to the rise of the First Dynasty in Egypt, including areas in which the goddess Neith was worshipped[13] and where the wearing of the tail of a fox was a common practice.[14] It is hard not to notice striking similarities to the Dogon in the Egyptian words and symbols attributed by Hagan to the Amazigh. These include the roots *akh*[15] and *paut,* which were symbolic in the Amazigh tradition of the "Eldest One of the Primeval Matter."[16] Even these few introductory references are highly suggestive of direct connections both to Dogon cosmology and to symbols and words of ancient Egypt.

In the collective name applied to the hunter tribes—the Amazigh—we also see suggestions of the Dogon. *Ama* would seem to refer to the name of the Dogon god Amma, and *zigh* calls to mind the Dogon Sigui festival and the Egyptian word *skhai,* meaning "celebration or festival." By this interpretation, the two parts of the name Amazigh would convey the meaning "celebrates Amma." This is just what one might expect to find if our placement of Dogon origins at the very beginning of Egyptian culture were a correct one. Such a title would be in keeping with the ancient designation for Egypt as the Land of Khem (completion), or based on Budge's definition of the word *khem* as meaning "Land of God."

During our discussions in the preceding chapters with their many hieroglyphic word examples, the very elegant nature of the Egyptian hieroglyphic language, its symbols, and its component structures cannot have escaped our notice. The features of this ancient language as interpreted here—its defined words and conceptual constructs—are highly suggestive of a deliberate, informed design. Such a suggestion should not surprise us, because it conforms to a known tradition within ancient Egypt itself that held—as did some other ancient cultures—that written language was a gift from ancestor/teacher gods. The many who will contend that such constructs can only be the product of wishful misinterpretation, subconscious manipulation, modern transference, or worse need only look to the explicit statements of the Dogon priests and the many direct corroborating references that date from the earliest days of Egyptian and Amazigh culture to see that these meanings must be native to the symbols themselves.

Throughout these studies, two observations have remained constant: first, that the events that the Dogon priests describe as Amma's efforts to create matter *do* accurately describe the component structures of matter; second, that the keywords and symbols used to define these structures *are* consistently reflected in the Egyptian hieroglyphic language. Based on these observations, we must reasonably conclude—in accordance with both Egyptian tradition and the archaeological record—that the Egyptian hieroglyphs constitute a designed system of language and that the design demonstrates an intimate familiarity with the processes by which modern scientists believe matter is formed.

The fundamental question, then, would seem to pertain to the identity of the ancestor/teachers themselves. To approach this question as we have many others in this study—through Egyptian language references—we begin with the Egyptian word *sba,*[17] meaning "to teach, to bring up, to educate, to instruct, to train." A related word meaning "pupil" or "teaching" is pronounced "sba-t"[18] and calls to mind the familiar Hebrew word for a day of instruction, *shabbat* (sabbath). Following this same line of inquiry, we see that these words serve almost as guide signs to what is perhaps a more significant word, *sbait,* meaning "teacher, instructor,"[19] and its homonym, *Sbait,* which Budge defines as a name for the star goddess Sothis, or Sirius. And so the words of the Egyptian language lead us back once again to one of the central controversies of Dogon mythology: the suggestion of a relationship between ancient knowledge and the stars of Sirius.

If Sirius had, in fact, played a pivotal role in the instructed civilization of humanity, wouldn't we expect to find something more than just a few symbolic words and mythological references directing us to the star? Shouldn't there be some revered monument or other obvious testament to the presumed great ancestor/teachers from Sirius? Ah, but if we only stop to consider carefully for a moment, we would realize that we are presented with just such monuments. Consider the three Great Pyramids of Giza, whose slightly off-center orientation seems to match the alignment of stars in the belt of Orion—the very stars that in a clear night sky point us in a directed line to Sirius itself. Also, how about the famous Ben Ben stone, the conical, perhaps meteoric stone that was said to have stood in Heliopolis, whose shape is re-created by the △ glyph—a glyph that Budge identifies with the dog star Septit?[20]

We are left to ponder how we should view this implied teacher/student relationship between Sirius and Earth. We might ask ourselves how, as an adult civilization, we should place the suggestion of such contacts in context? Again, the indication is that we should follow myth and the Egyptian language and heed the four-stage symbolic progression they describe—from insects to fish to four-legged canids to flying birds. In the Dogon tradition, Earth is referred to as the fish star whereas Sirius is considered to be the dog star, as it also was in ancient Egypt. If we are to think of the advancement of civilization as a similar kind of four-stage progression, then these ancient appellations may have been meant to reflect our relative place—and that of our teachers—within that progression.

Although the trail of the Egyptian language seems to begin sometime after 3400 BCE, there are persistent signs that the Sphinx could actually be much older. These hints, taken in the context of many Egyptian references to a revered "First Time," suggest that there may yet be another earlier era of human history undiscovered and still undocumented.

AFTERWORD

As a consequence of my research for *The Science of the Dogon* and *Sacred Symbols of the Dogon,* I had long presumed that there were underlying similarities between the cosmological symbols of India and those of Egypt and the Dogon. However, it was not until my daughter Hannah returned from a visit to India with stories of a traditional ritual shrine called the *chorten* or *stupa* that I realized the depth of the resemblance. The stupa can take many different forms, but its classic plan appears to be the very image of a diagram from Marcel Griaule's *Conversations with Ogotemmeli*—that of the Dogon granary. I suggested to my daughter that I believed I could blindly predict the traditional symbolism of the stupa based simply on my knowledge of the Dogon granary. I then recounted to her some of the ways in which the granary is said to represent the plan of a world system, incorporating many fundamental geometric shapes and representing key structures of the solar system and of cosmology. I then did some basic online research relating to the stupa and discovered that the references supported each of my symbolic claims for the stupa based on the Dogon pattern.

Hoping to learn more, I ordered a copy of a book titled *The Symbolism of the Stupa* by Adrian Snodgrass, an internationally renowned authority on Buddhist symbolism, architecture, and art and author of classic works on Buddhist symbolism. In the book, Snodgrass describes the stupa as the preeminent symbol of the Buddhist concept of *dharma,* provides an exhaustive study of its role as the keystone of a symbolic world plan, and traces many intimate details of its significance within widespread religious traditions of India and Asia.

As Snodgrass describes it, the ground plan of the stupa closely matches that of the Dogon granary. Each structure re-creates the form of the Dogon egg-in-a-ball drawing—the figure in Dogon cosmology that defines how multiplicity is manifested from unity. Snodgrass walks us through the geometry of the construction of the stupa and, in the process, provides supporting detail for Griaule's description of the Dogon granary (see the figure on page 36). He also—perhaps unknowingly—demonstrates how the shapes and cosmological meanings of several key Egyptian glyphs can be derived from the stupa's form.

The construction of the stupa begins with an unordered plot of land and a stick. The stick is used as a gnomon (the central vertical piece of a sundial whose shadow is cast in the sunlight). It is placed vertically in the ground and becomes the midpoint of a circle with a radius twice the length of the height of the stick. During the day, the two longest long shadows cast by the stick—one in the morning and one in the afternoon or evening—will intersect the circle at two different points. Because of the relative positions of these points, any line drawn between them will be intrinsically oriented along an east/west axis. The line itself will only pass through the gnomon on two specific days of the year—on the dates of the two equinoxes. On all other days, the line's position will move progressively farther away from the gnomon until the date of the next solstice, when it will begin to move back toward the gnomon. So the basic structure of the stupa, which is derived from a circle with a central point ☉, serves both as a sundial to track the hours of the day and as a kind of observational tool to track the long-term movements of the sun, and through them to track the seasons and years. Thus, the first stage in the defined form of the stupa involves (based on the rotation of the earth in relation to the sun and the orbit of Earth around the sun) both the Egyptian sun glyph shape and its traditional meanings in the Egyptian language—representing all at once the sun, a day, and a period of time.

According to Snodgrass, the initiate who constructs the stupa next draws two more circles that are centered on the two points of intersection of the gnomon's shadow with the original circle, each with a radius somewhat larger than that of the original circle. These two new circles will intersect each other at two points, and a line drawn between these points will define an oriented north/south axis that will pass through the gnomon.

A few more arcs are then drawn that allow the initiate to define the

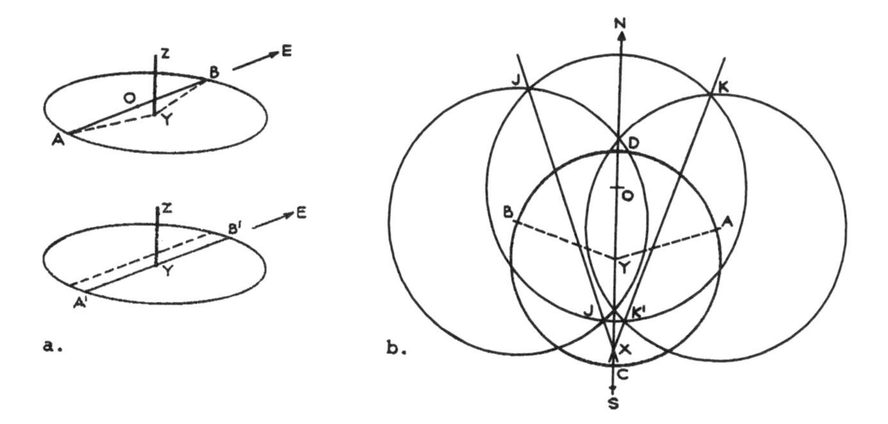

The determination of the orient
(from Snodgrass, The Symbolism of the Stupa, *15)*

corner points of a square based on the same central point as the original circle. This square □ is considered to be symbolic of an ordered space that conceptually has been manifested by the stupa from the chaos of the previously undefined plot of land. Thus, the plan of the stupa also produces an image that matches our Egyptian glyph for space and its cosmological meaning in much the same way that it previously produced the sun glyph and its symbolic meanings.

Although the symbolism of the stupa as Snodgrass relates it is very extensive and often esoteric in nature, he describes a system that he feels is intended, like Dogon cosmology, to define, all at once, the

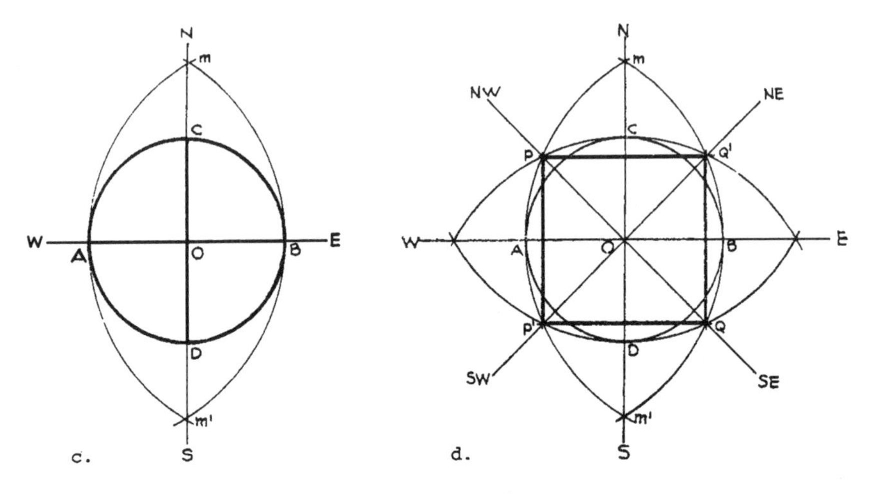

Further determination of the orients
(from Snodgrass, The Symbolism of the Stupa, *16)*

emergence of matter on a microscopic level, the emergence of the universe on a macroscopic level, and the emergence of mankind on both a biological and spiritual level. This system begins with the formless and timeless unity of the primordial waters, is characterized by imagery of growing plants, is facilitated by cords conceived as rays that emerge in increments of seven, and culminates in a divine Word. In many ways, the symbolism of the stupa as Snodgrass describes it serves as a kind of bridge between some of the seemingly disparate aspects of Indian, Dogon, and Egyptian cosmology. For instance, in the Egyptian system, it is the mother goddess Neith who weaves matter on a loom with her shuttle, and in the Dogon system, the spider Dada (whose name means "mother") weaves the threads of matter. Snodgrass tells us that in Buddhist symbolism, matter is woven by a spider in a manner similar to that of a shuttle on a loom, that the web of the spider is considered to be a spiraling coil converging on its center ᘓ, and that the web itself is comprised of seven rays.

The mythological episodes that Snodgrass relates call to mind specific tales of the Dogon and the Egyptians, often serving to reconcile some of their differences in symbolism. Although the Buddhist mythological keywords from India as noted by Snodgrass do not arise from root sounds that seem obviously related to corresponding Dogon and Egyptian keywords, those words do express many of the same concepts and define them in the same fundamental terms and so are almost entirely recognizable to a reader who is familiar with the Dogon and Egyptian cosmological systems. In the end, *The Symbolism of the Stupa* confirms symbol after symbol as documented among the Dogon by Marcel Griaule, defines a cosmological system that runs precisely parallel to the Dogon and Egyptian systems, and seems almost entirely intelligible within their context.

If we are to believe what Dogon and Egyptian cosmology and language tell us and what Snodgrass explicitly confirms in terms of Buddhist symbolism, then this shared mythological concept of matter and how it is formed carries some potentially significant implications for modern scientists. For the purposes of this discussion, we should think of time and mass as a continuum. At one extreme end of this continuum we can imagine underlying waves, which we believe exist in an immensely fast time frame and carry virtually no mass or acceleration; at the other end we can imagine infinite acceleration or infinite mass, which would exist in an immensely slow time frame.

Between these extremes we have the processes of the formation of mass, governed by the Calabi-Yau space. Our understanding of this process begins with a Dogon metaphor for the translation of waves into particles—the tradition of preserving beans by covering them in sand and then employing a sieve to separate them from the sand before use. Within the mind-set of this metaphor, if the beans represent particles and the sand represents waves, then the Egyptian language tells us that the Calabi-Yau space, which is the structure through which mass is woven and that the Dogon characterize as a spiraling coil, must be represented by the sieve, as shown in the following table:

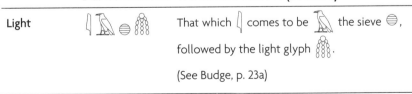

DEFINITION OF THE COILED THREAD GLYPH (SKHET)

To weave		The bending sieve of mass, followed by the coiled thread glyph. (See Budge, pp. 694b, 695a)

We also know based on an ideographic reading of the Egyptian word *aether* that time itself must be the elusive fabric of the universe from which mass is woven.

THE BENDING FORCE CAUSES AN ORBIT (AETHER)

An orbit/ unit of time	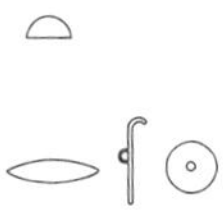	Mass bends/warps time followed by sun glyph. Budge interprets this word to mean "time or season." (See Budge, p. 101b)

These conceptualizations make complete scientific sense because Albert Einstein tells us that an increase in mass slows or decreases the time frame. It is as if through this process of creation time is converted into mass. However, what does the term *immensely fast* mean in terms of time? Einstein tells us that within each relative time frame the fundamental "speed limit," so to speak, is the speed of light. So let's presume that massless waves—even within their own ultrafast time frame—cannot be woven from waves into particles at a rate faster than the speed of light. We can see evidence of this perspective expressed in the following alternate reading of the Egyptian word *aakhu,* meaning "light."

DEFINITION OF THE LIGHT GLYPH (AAKHU)

Light	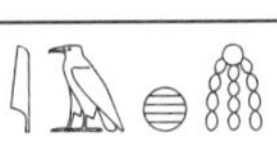	That which comes to be the sieve, followed by the light glyph. (See Budge, p. 23a)

So, essentially, massless waves, which in effect are time and which constitute the fabric of the universe, enter the bending sieve of the Calabi-Yau space, which is light, and vibrate, causing what ultimately appear to us as the various subatomic particles of matter. Effectively, these threads (as the Dogon refer to them)—the fabric of the universe—"point" toward the woven particles in the same way that a string of yarn coming out of a rolled ball of yarn "points" toward a knitted sweater. The greater the mass of the object, the greater the number of particles that must comprise it, and therefore the more threads that ultimately must "point" back toward that object. So the formation of the body comes to bend the fabric of time toward itself in much the same way as described by Einstein. Likewise, if the associated Calabi-Yau spaces are strung together as Buddhist tradition describes, "like pearls on a string," then the greater the mass of the body, the longer the circuitous detour becomes that the waves (which are time) must make (at the speed of light) through this trail of intervening coils to traverse the body, and so time slows down in respect to that body. This perspective is also supported by various Egyptian words for *time*. The first qualifies as a defining word for the time glyph and is written as follows:

MASS BENDS TIME (ATRU)

Definition of time		That which mass bends , followed by the time glyph . Budge interprets this word to mean "time or season." (See Budge, p. 100a)

Another Egyptian word for *time* can be seen as an alternate defining word for the sun glyph, our symbol for the major effect of gravity, an orbit, which would be caused by the bending of time.

THE BENDING OF TIME CAUSES AN ORBIT (TERA)

An orbit/ unit of time		Bent time, followed by the sun glyph. Budge interprets this word to mean "time or season." (See Budge, p. 884b)

Perhaps the final point to be made in regard to the Egyptian hieroglyphs is illustrated by the above examples. It is that the ideographic meanings of Egyptian words as we understand them cross-check to each other conceptually. The glyphs of Egyptian words that, according to Budge, carry related traditional meanings express ideographic concepts that are also clearly related to each other. This consistency within the language itself constitutes yet one more affirmation of its designed nature and of the correctness of the interpreted glyph meanings presented in this book.

APPENDIX A

EGYPTIAN GLYPHS BY CONCEPT

LIST OF EGYPTIAN GLYPHS—ALPHABETIC BY CONCEPT

Glyph	Ideographic Meaning
	Adze; symbol of the formation of mass from waves
	Arc, bend
	Atom
	To become, to come to be
	To behold (to know, known)
	To bend, to enclose
	To bend and grow
	Bending force (gravity)
	To bind
	To break, to pierce
	Building; structure; rotate

LIST OF EGYPTIAN GLYPHS—ALPHABETIC BY CONCEPT *(cont.)*

Glyph	Ideographic Meaning
	Calabi-Yau space begun
	Calabi-Yau space complete
	Calabi-Yau space complete; to break, to bend
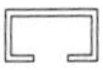	Chamber; pylon; wrapped-up dimension of string theory
	Coiled thread (seven vibrations complete)
	Coiled thread grows
	To come to be, to become
	To conceal, to hide
	Creative force
	Delimitation posts; to stand up
	DNA, double helix
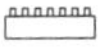	Drawing/weaving force; to draw; to weave
	Duality; by
	Earth, mass, matter, mass complete
	Earth, mass
	Earth, mass
	Earth bent or bound
	Effect; limit; source
	Effect of gravity

Glyph	Ideographic Meaning
	Egg; womb of all world signs
	Electron orbit (from Dogon cosmology)
	Emanation (one of the "twin doors" of the Dogon cosmology)
	To encircle, to orbit
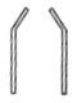	To enclose, to bend
	Event horizon of a black hole; shape of the unformed universe
	Force (a force); to make, to cause
	Force of; strength of; a piece or portion of; to give; to transport
	To give; to make
	Gravity, bending force
	To grow and bend
	Growth comes to be
	Growth and vibration come to be
	Growth of mass from waves
	To hide, to conceal
	Knot (cause of particles in torsion theory)
	To know, known; to behold
	Letters; learning
	Light

LIST OF EGYPTIAN GLYPHS—ALPHABETIC BY CONCEPT *(cont.)*

Glyph	Ideographic Meaning
	Limit; effect; source
	To make; to give
	Mass, matter, mass complete (earth)
	Mass, earth
	Mass, earth
	Mass, matter, earth
	Mass, matter
	Mass, matter
	Mass complete
	Mass increases
	Mass increased
	Mass raised up
	Mass raised up
	Massless wave
	Matter, mass, mass complete (earth)
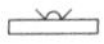	Membrane rolled; completed
	Moon
	Nonexistence
	To orbit, to encircle

Glyph	Ideographic Meaning
	Orbit of Earth around the sun; the sun; day; period of time
	Origin
	Particle
	Pedestal
	Perception, to see
	Perception
	Perception draws waves up (fire)
	Permanent, established
	To pierce, to stab
	Place (smaller places); to grow
	Primeval
	Proton, neutron, or electron
	Quark
	To raise up
	To rotate; building; structure
	To see; perception
	Seeds
	Seeds or signs, the three groups: guide signs, master signs, and world signs

LIST OF EGYPTIAN GLYPHS—ALPHABETIC BY CONCEPT *(cont.)*

Glyph	Ideographic Meaning
	To set, to set in place
	Source; limit; effect
	Space
	To speak, to utter a word or a sound
	To stand up; delimitation posts
	String intersection (self-intersection of a string)
	String intersection (intersection of two strings)
	String intersection (complex intersection of strings)
	Strong nuclear force (Dogon "stocky")
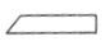	Subtraction
	That which; that which is
	Time
	To transmit
	Tree or wood (that which fire burns)
	To vibrate
	Vibration (wind)
	Vibrations
	Vibrations as rays of a star; the Tuat
	Vibration and growth come to be

Glyph	Ideographic Meaning
	Wave
	To waver
	Waves
	Waves draw up
	Wave force; the electromagnetic force (Dogon "bumpy")
	Weak nuclear force (Dogon "that bows its head")
	To weave
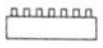	To weave; drawing/weaving force; to draw
	Wind; vibration
	Word (membranes in string theory)
	Word given
	Writing

APPENDIX B

IDEOGRAPHIC WORD EXAMPLES

DEFINITION OF THE LIGHT GLYPH (AAKHU)

Light	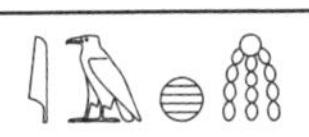	That which comes to be the limit , followed by the light glyph . (See Budge, p. 23a)

DEFINITION OF THE SUN GLYPH (AAT-T)

Hour, time		Vibrations of mass , followed by the sun glyph , referring to the sun glyph's symbolism as representing a unit of time. (See Budge, p. 27a)

DEFINITION OF A MONTH (ABT)

Month		The moon makes an orbit . (See Budge, p. 40b)

THE BENDING FORCE CAUSES AN ORBIT (AETHER)

An orbit/ unit of time		Mass bends/warps time , followed by the sun glyph . Budge interprets this word to mean "time, or season." (See Budge, p. 101b)

DEFINITION OF A COURTYARD (AHI)

Courtyard		That which encircles growth , followed by the courtyard glyph. (See Budge, p. 74b)

DEFINITION OF KNOWLEDGE (AM)

To know, to understand		The act of perceiving that which is the coiled thread through word written , followed by the looped tie glyph (used to tie a scroll). (See Budge, p. 120a)

DEFINITION OF THE EYE GLYPH (AN)

Perception		The force or act by which waves attain mass via the coiled thread , followed by the eye glyph . (See Budge, p. 123a)

DEFINITION OF THE ADZE GLYPH (AN-T)

Adze

The act or force by which waves attain mass, followed by the adze glyph.
(See Budge, p. 123b)

DEFINITION OF A DOORKEEPER (ARIT-AA)

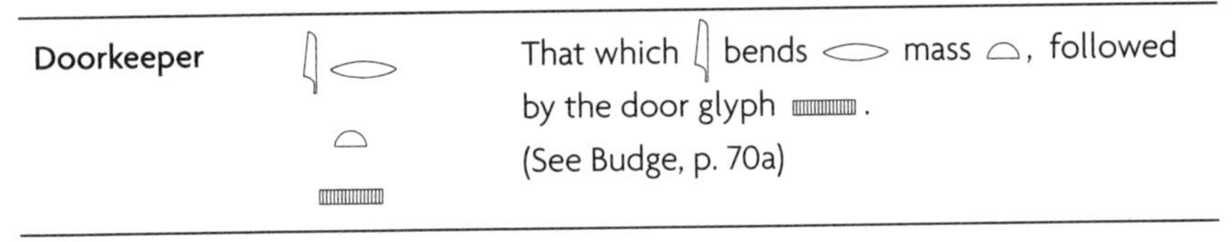

Doorkeeper

That which bends mass, followed by the door glyph.
(See Budge, p. 70a)

DEFINITION OF THE SERPENT GLYPH (ATCHET)

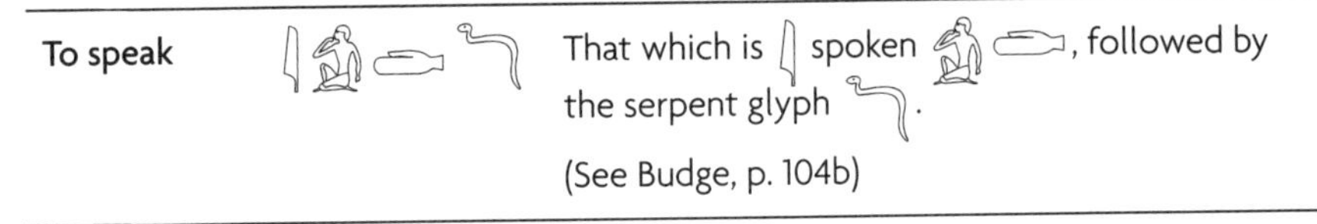

To speak

That which is spoken, followed by the serpent glyph.
(See Budge, p. 104b)

DEFINITION OF THE KNOT GLYPH (ATENNU)

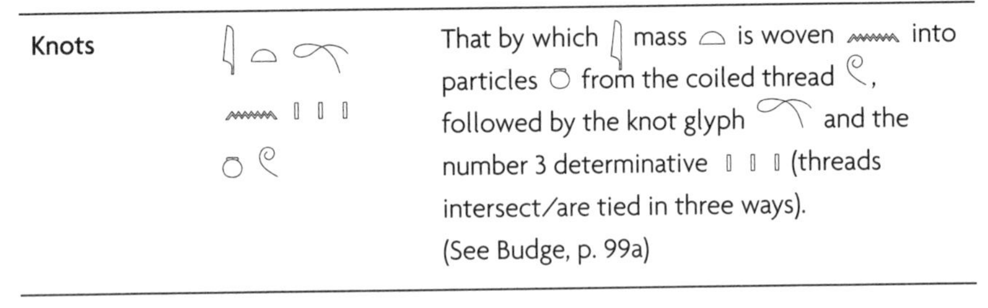

Knots

That by which mass is woven into particles from the coiled thread, followed by the knot glyph and the number 3 determinative (threads intersect/are tied in three ways).
(See Budge, p. 99a)

DEFINITION OF THE TIME GLYPH (ATRU)

Time		That which mass bends or warps , followed by the time glyph . (See Budge, p. 100a)

DEFINITION OF THE NEST GLYPH (AUN)

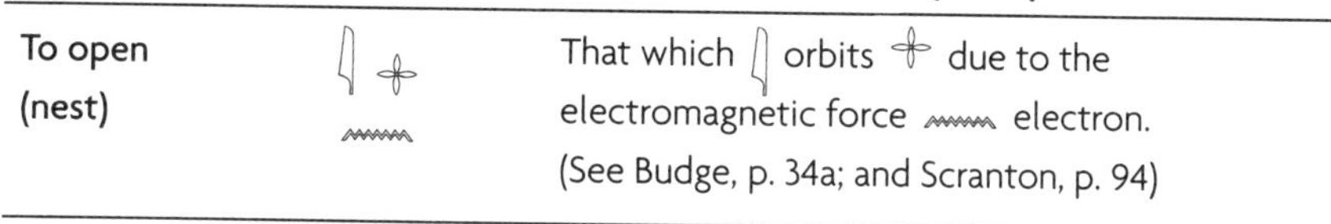

To open (nest)		That which orbits due to the electromagnetic force electron. (See Budge, p. 34a; and Scranton, p. 94)

DEFINITION OF THE CURVED STAFF GLYPH (AU-T)

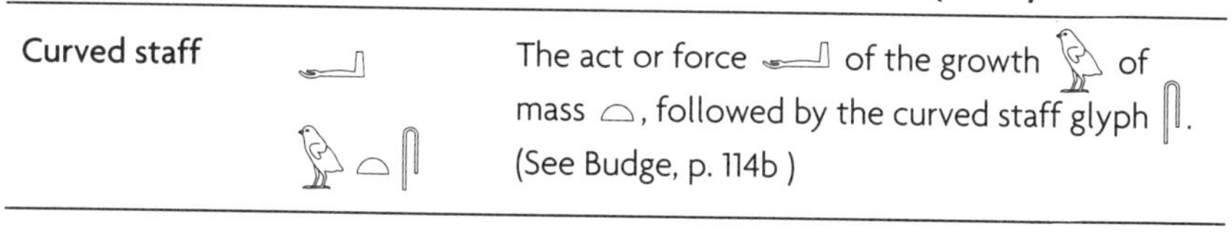

Curved staff		The act or force of the growth of mass , followed by the curved staff glyph . (See Budge, p. 114b)

PLACE OF TRUTH (BU MAA)

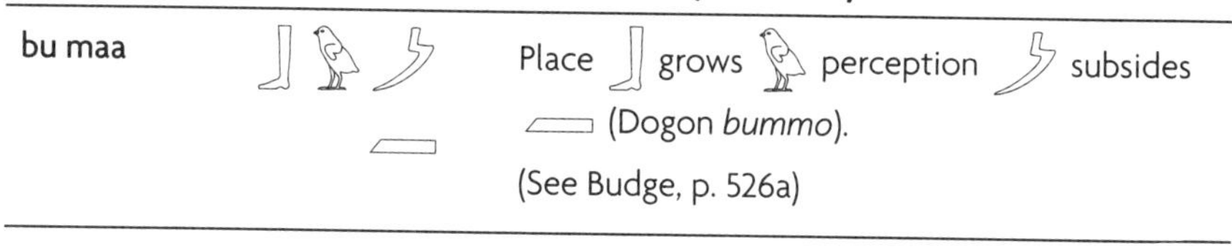

bu maa		Place grows perception subsides (Dogon *bummo*). (See Budge, p. 526a)

DEFINITION OF GOD (KHEM)

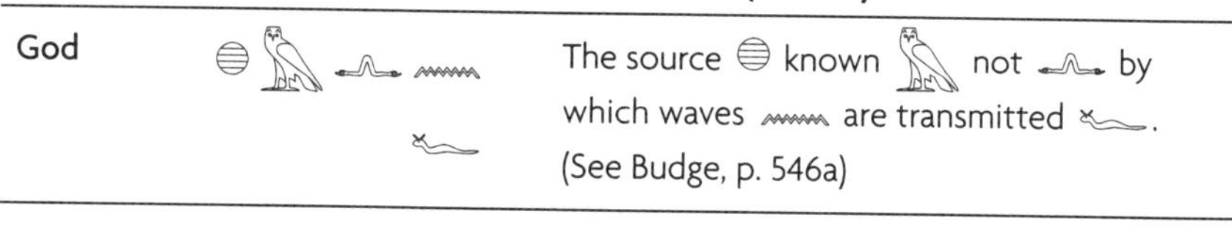

God		The source known not by which waves are transmitted . (See Budge, p. 546a)

DEFINITION OF THE SCARAB GLYPH IN ITS ASPECT AS NONEXISTENCE (KHEPER)

Nonexistence		Not , created , followed by the scarab glyph . (See Budge, p. 542a)

DEFINITION OF THE SCARAB GLYPH IN ITS ASPECT OF COMING INTO EXISTENCE (KHEPER)

To come into existence	□	Source of space □, followed by the scarab glyph . (See Budge, p. 542a)

DEFINITION OF THE FALLING MAN GLYPH (KHER)

To fall down		Result/product/effect of the bending/warping force followed by the falling man glyph (effect of gravity). (See Budge, p. 560b; and the word *sher,* p. 749b)

GOD OF THINGS THAT EXIST (KHET)

The god Khet		Source of mass or matter , followed by the number 3 determinative and the god determinative . (See Budge, p. 526a)

DEFINITION OF COMPLEX STRING INTERSECTION GLYPH (NET)

Name of Neith		Weaves matter , followed by the complex string intersection and the goddess glyph determinative . (See Budge, p. 399b)

DEFINITION OF A GOD (NETHER)

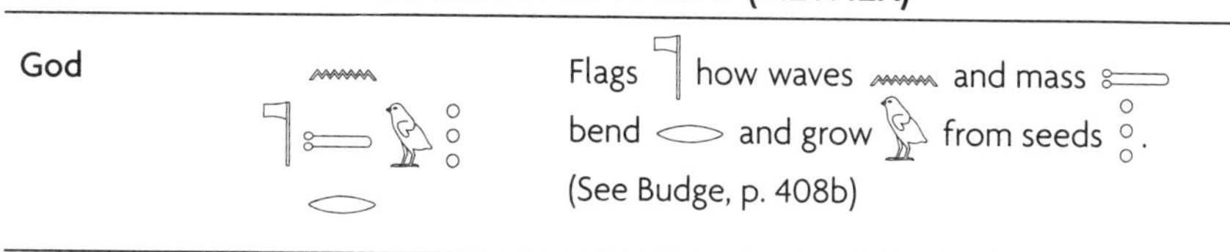

God		Flags how waves and mass bend and grow from seeds . (See Budge, p. 408b)

DEFINITION OF THE SIMPLE STRING INTERSECTION GLYPHS (NT-T)

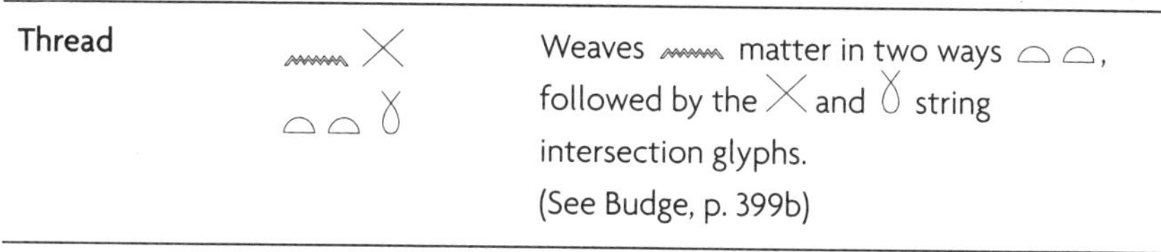

Thread		Weaves matter in two ways , followed by the and string intersection glyphs. (See Budge, p. 399b)

WAVES GROW (NU)

Plural of the wave glyph		Waves grow . (See Budge, p. 349a)

AN ADZE (NU)

Adze		Forms/shapes particles from coiled threads . (See Budge, p. 352a)

DEFINITION OF THE LOOPED STRING INTERSECTION GLYPH (NU)

To tie, to bind together		Waves form/shape particles from coiled threads , followed by the looped string intersection. (See Budge, p. 351a)

DEFINITION OF HEMISPHERE GLYPH (PAU-T)

Mass, matter	Existence primeval (mass or matter), followed by the hemisphere glyph . (See Budge, p. 230b, "stuff, matter, substance, material of which anything is made," the penultimate spelling)

THE CONCEPT OF A VORTEX (PEKHAR-PEKHAR)

Vortex	The circuit of the spiraling coil's flow . (See Budge, p. 247a)

WHAT COMES FORTH FROM THE MOUTH (PER)

Word, speech	The structure of bending spoken . (See Budge, p. 240b)

DEFINITION OF MANIFESTATION (PER-T)

Manifestation (To go out, to go forth)	The coming forth of gravity , mass , and the coiled thread . (See Budge, p. 240b)

DEFINITION OF THE CHAMBER GLYPH (PER-TUAT)

Chamber in the other world	Chamber gives growth to vibrations of mass , followed by the chamber glyph . (See Budge, p. 240a)

ARCHITECT OF HEAVEN, EARTH, AND THE BODIES OF MEN (PTEH)

The god Pteh		Space, mass/womb, and DNA, followed by the god determinative. (See Budge, p. 254b)

DEFINITION OF THE SUN GLYPH (RA)

Sun glyph		The bending/warping force, followed by the sun glyph —the image of an orbit. (See Budge, p. 417b)

DEFINITION OF THE SUN GLYPH (RA)

The sun, the day		The bending/warping force, followed by the orbit glyph, and the number 1 determinative. (See Budge, p. 417b)

ENUMERATING DEFINITION OF THE BENT ARM GLYPH (REMEN)

A force		Bending force, drawing/weaving force, wave force, followed by the bent arm glyph and the number 3 determinative. (See Budge, p. 425a)

DEFINITION OF A YEAR (RENP-T)

Year		The time of Earth's orbit around the sun. (See Budge, p. 427b)

DEFINITION OF THE TIME GLYPH (SAB)

Time, period		Binds space , followed by the time glyph . (See Budge, p. 588b)

DEFINITION OF THE WEAK NUCLEAR FORCE GLYPH (SEN)

To bow or pay homage		The binding force followed by the weak nuclear force glyph . (See Budge, p. 603a)

PROTON/NEUTRON (SEN/SENE)

They, them, their		The binding of a particle from three elements (Protons and neutrons are each formed from different combinations of three quarks). (See Budge, p. 603a; Definition of a proton/neutron [sen/sene]; and Scranton, p. 94)

DEFINITION OF THE STRONG NUCLEAR FORCE GLYPH (SENNU)

Image		The binding force of forces followed by the strong nuclear force glyph . (See Budge, p. 604b)

DEFINITION OF A PARTICLE (SENU)

Pot, vase		Bound particle . (See Budge, p. 605b)

DEFINITION OF SEASONS (SES)

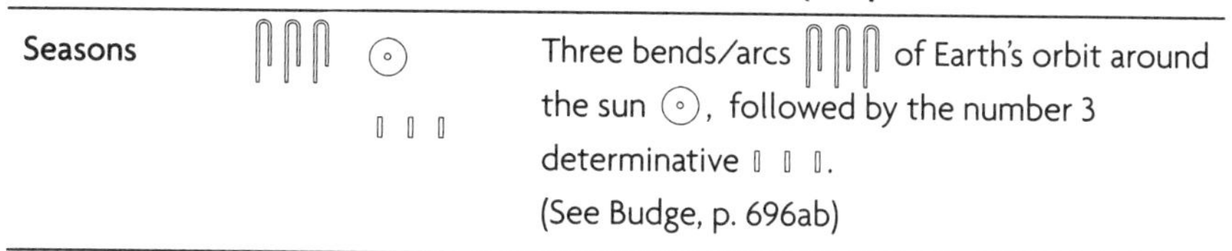

Seasons	Three bends/arcs of Earth's orbit around the sun, followed by the number 3 determinative. (See Budge, p. 696ab)

DEFINITION OF THE STABBING/PIERCING GLYPH (SET)

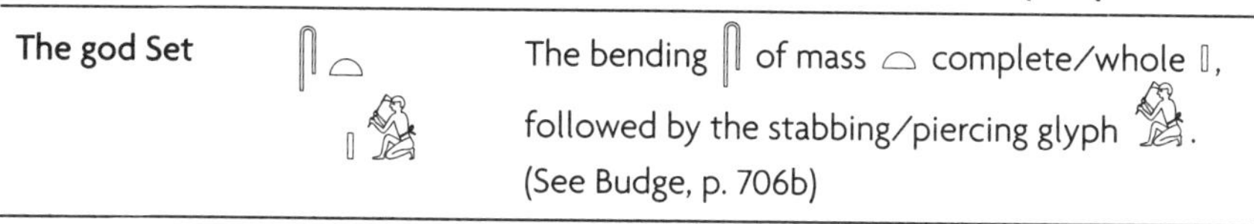

The god Set	The bending of mass complete/whole, followed by the stabbing/piercing glyph. (See Budge, p. 706b)

DEFINITION OF THE STAR GLYPH (SET)

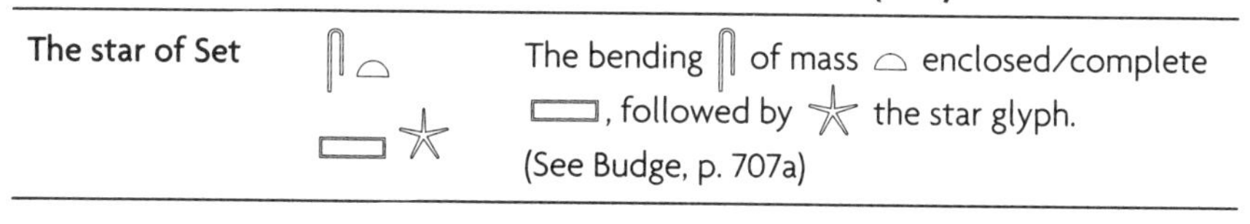

The star of Set	The bending of mass enclosed/complete, followed by the star glyph. (See Budge, p. 707a)

DEFINITION OF ACCELERATION (S-KHAKH)

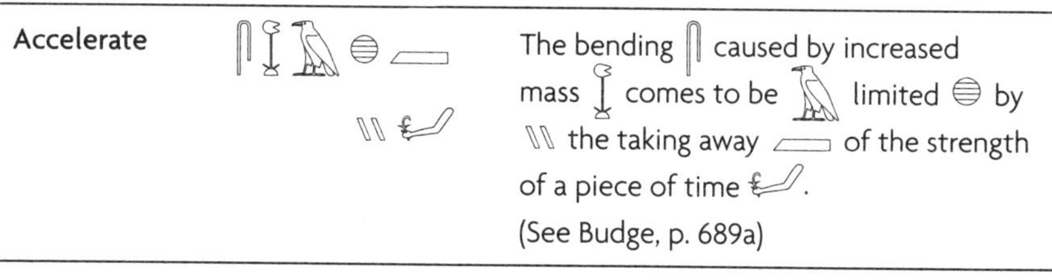

Accelerate	The bending caused by increased mass comes to be limited by the taking away of the strength of a piece of time. (See Budge, p. 689a)

DEFINITION OF THE COILED THREAD GLYPH (SKHET)

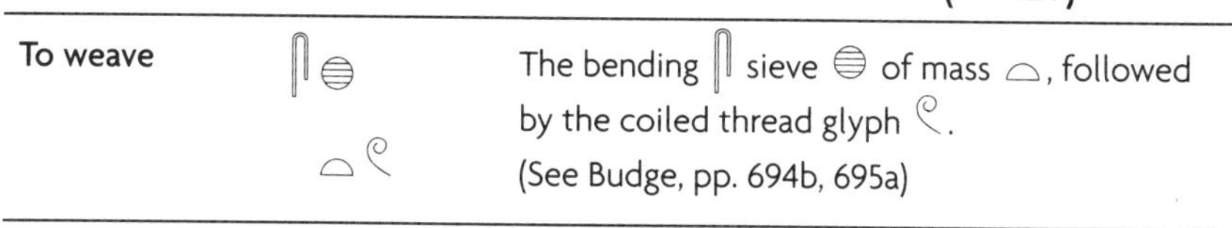

To weave	The bending sieve of mass, followed by the coiled thread glyph. (See Budge, pp. 694b, 695a)

DEFINITION OF THE COURTYARD GLYPH (TA-T)

Room, chamber		The drawing up and encirclement of mass, followed by the courtyard glyph. (See Budge, p. 821b)

DEFINITION OF THE FALCON GLYPH (TCHET)

The Divine Word		Serpent given growth, followed by by the falcon glyph/god determinative. (See Budge, p. 913b)

DEFINITION OF THE COURTYARD GLYPH (TENNU)

Border, boundary		Gives waves existence, followed by the courtyard glyph and the numbers 1 and 3 determinatives. (See Budge, p. 881b)

THE BENDING OF TIME CAUSES AN ORBIT (TERA)

An orbit/ unit of time		Bent time, followed by the sun glyph. (See Budge, p. 884b) Budge interprets this word to mean "time or season."

DEFINITION OF THE LOOP STRING INTERSECTION GLYPH (TU)

Bandlet		Made from the coiled thread, followed by the looped string intersection glyph. (See Budge, p. 870a)

DEFINITION OF THE TOWN GLYPH (TUA-T)

Ancient name for the Tuat		Gives growth to mass, followed by the town glyph. (See Budge, p. 871b)

DEFINITION OF THE PYLON/CHAMBER GLYPH (TUA-T)

Ancient name for the Tuat		Gives growth that becomes mass, followed by the pylon glyph. (See Budge, p. 871b)

DEFINITION OF THE PYLON GLYPH (TUA-T)

Ancient name for the Tuat		Vibrations in the Calabi-Yau space become mass, followed by the pylon glyph. (See Budge, p. 871b)

DEFINITION OF THE STAR GLYPH (TUAIT)

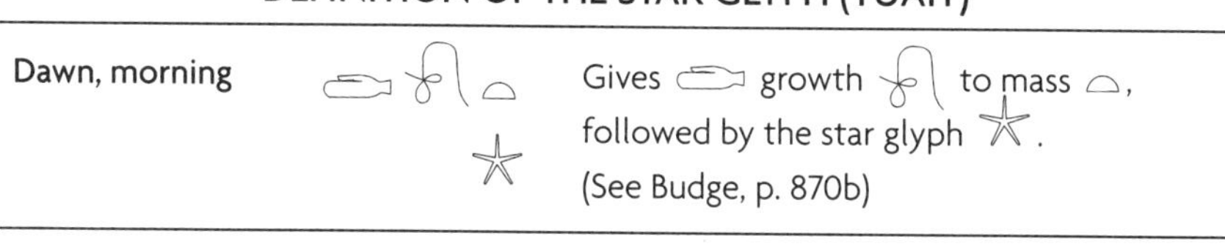

Dawn, morning		Gives growth to mass, followed by the star glyph. (See Budge, p. 870b)

THE BIRTH GODDESS (UA)

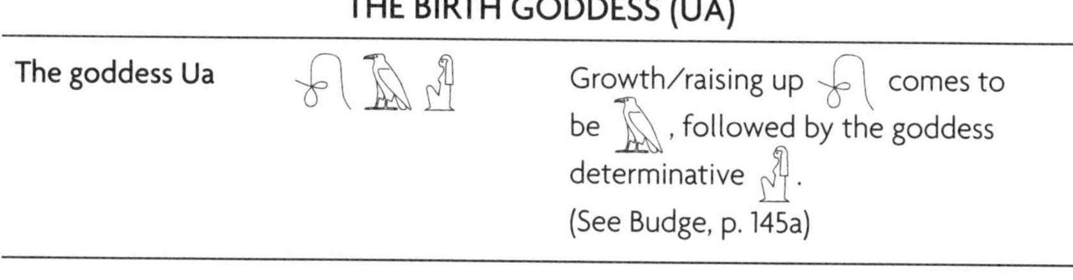

The goddess Ua		Growth/raising up comes to be, followed by the goddess determinative. (See Budge, p. 145a)

DEFINITION OF THE COILED THREAD GLYPH (UN-T)

Rope, cord		Vibration creates mass , followed by the coiled thread glyph . (See Budge, p. 167a)

DEFINITION OF THE STAR GLYPH (UNTI)

Light god, god of an hour	//	Vibration causes matter to be dual //, followed by the star glyph . (See Budge, p. 167b)

DEFINITION OF THE PIERCING/STABBING GLYPH (UNTI)

Opener, piercer, stabber	//	Vibration causes matter to be dual //, followed by the piercing/stabbing glyph . (See Budge, p. 166b)

DEFINITION OF THE GRAVITY GLYPH (UNU-T)

Hour, moment	//	That which becomes given to mass when it is divided in two //, followed by the sun glyph and the number 1 determinative , referring to the sun glyph's symbolism as representing a unit of time. (See Budge, p. 167a)

DEFINITION OF THE STAR GLYPH (UNU-T)

Hour, time		Vibrations of particles of mass , followed by the star glyph . (See Budge, p. 167a)

NOTES

Chapter 1: Dogon Science in the Egyptian Hieroglyphs

1. Laird Scranton, *The Science of the Dogon: Decoding the African Mystery Tradition* (Rochester, Vt.: Inner Traditions, 2006), 98–100.
2. E. A. Wallis Budge, *An Egyptian Hieroglyphic Dictionary* (New York: Dover Publications, 1978), 51b, 54a.
3. Marcel Griaule and Germaine Dieterlen, *The Pale Fox* (Chino Valley, Ariz.: Continuum Foundation, 1986), 33.

Chapter 2: Science and the Structure of Matter

1. Stephen W. Hawking, *A Brief History of Time* (Toronto: Bantam, 1988), 87.

Chapter 3: Dogon Cosmology

1. Scranton, *The Science of the Dogon,* 107.
2. Ibid., 69.
3. Ibid., 78.
4. Ibid., 77.
5. Ibid., 35–37.

Chapter 4: Dogon Symbols and Egyptian Glyphs

1. Budge, *An Egyptian Hieroglyphic Dictionary,* 106b.
2. James P. Allen, *Middle Egyptian: An Introduction to the Language and Culture of Hieroglyphics* (Cambridge: Cambridge University Press, 2000), 459.

3. Budge, *An Egyptian Hieroglyphic Dictionary,* 15a.
4. Scranton, *The Science of the Dogon,* 99.

Chapter 5: Defining Egyptian Glyphs

1. Budge, *An Egyptian Hieroglyphic Dictionary,* 351a; and Scranton, *The Science of the Dogon,* 107.
2. Scranton, *The Science of the Dogon,* chap. 7.
3. Budge, *An Egyptian Hieroglyphic Dictionary,* 230b.
4. Scranton, *The Science of the Dogon,* 93.
5. Budge, *An Egyptian Hieroglyphic Dictionary,* 230b.
6. Ibid., cxiv.
7. Griaule and Dieterlen, *The Pale Fox,* chap. 2.
8. Budge, *An Egyptian Hieroglyphic Dictionary,* 694b, 695a.
9. Ibid., cxlv.
10. Ibid., 247a.
11. Ibid., 99a.
12. Ibid., 114b.
13. Ibid., 197a.
14. Ibid., 421b.

Chapter 6: Egyptian Concepts of Astrophysics

1. Genevieve Calame-Griaule, *Dictionnaire Dogon* (Paris: Librarie C. Klincksieck, 1968), 205.
2. Griaule and Dieterlen, *The Pale Fox,* 156.
3. Budge, *An Egyptian Hieroglyphic Dictionary,* 525a.
4. Ibid., 525a.
5. Ibid., 529a.
6. Ibid., 541a.
7. Ibid., 544b.
8. Ibid., 526b.
9. Ibid., 255b.
10. Ibid., 743b.
11. Ibid., 780a.
12. Ibid., cxxi.
13. Scranton, *The Science of the Dogon,* 67.
14. Ibid., 78.
15. Budge, *An Egyptian Hieroglyphic Dictionary,* 40b.

16. Ibid., 696a.
17. Ibid., 588b.
18. Ibid., 105a.
19. Scranton, *The Science of the Dogon,* 78.
20. Budge, *An Egyptian Hieroglyphic Dictionary,* 34a.
21. Ibid., cxxiii.
22. Ibid., 603a.

Chapter 7: Egyptian Glyphs, Words, and Deities

1. Budge, *An Egyptian Hieroglyphic Dictionary,* cxvi.
2. Ibid., 230b.
3. Ibid., 231a.
4. Ibid., cxxvi.
5. Ibid., 605b.
6. Ibid., 349a.
7. Ibid., 349b.
8. Ibid., cxix.
9. Ibid., 913b.
10. Scranton, *The Science of the Dogon,* 71.

Chapter 8: The Nummo Fish

1. Griaule and Dieterlen, *The Pale Fox,* 184.
2. Budge, *An Egyptian Hieroglyphic Dictionary,* 266a.
3. Ibid., cxxvi, entry 56.
4. Ibid., 121b.
5. Ibid., 821b.
6. Ibid.
7. Ibid.
8. Ibid., 821a.
9. Ibid.
10. Ibid. 815–21.
11. Ibid., cxxix.
12. Ibid., 779b.
13. Ibid., 872a.
14. Ibid.
15. Griaule and Dieterlen, *The Pale Fox,* 186.
16. Budge, *An Egyptian Hieroglyphic Dictionary,* 182a–182b.

17. Ibid., 180a–180b.
18. Ibid., 146a.
19. Ibid., 148a.
20. Scranton, *The Science of the Dogon,* 77; and Griaule and Dieterlen, *The Pale Fox,* 137–38.
21. Budge, *An Egyptian Hieroglyphic Dictionary,* 913b.

Chapter 9: Symbolic Structure of the Egyptian Language

1. Budge, *An Egyptian Hieroglyphic Dictionary*, 123a.
2. Ibid., cxx.
3. Ibid., 121b.
4. Ibid.
5. Ibid., 121a.
6. Ibid., 120a.
7. Ibid., 191b.
8. Ibid., 267a.
9. Ibid., 266b.
10. Ibid., 268b.
11. Ibid., 272b.
12. Ibid., 152a.
13. Ibid.
14. Ibid., 153a.
15. Ibid., 151a.
16. Ibid., 525a.
17. Ibid., 526a.
18. Ibid., 566b.
19. Ibid., 779b.
20. Ibid., 780b.

Chapter 10: The Tuat

1. Budge, *An Egyptian Hieroglyphic Dictionary,* 588a.
2. Ibid.
3. Ibid., 871a.
4. Ibid., 870b.
5. Ibid., 872a.
6. Scranton, *The Science of the Dogon,* 80.
7. Budge, *An Egyptian Hieroglyphic Dictionary,* 868b.

8. Ibid., 870b.
9. Ibid., 871b.
10. Ibid., 167a.
11. Ibid., 167a.
12. Ibid., 169b.
13. Ibid., 174b.
14. Ibid., 174a, 171b.
15. Ibid., 130b.
16. Ibid., 432a.
17. Ibid., 432b.
18. Ibid., 176b.
19. Ibid., 177a.
20. Ibid.
21. Ibid., 167b.
22. Ibid., 166b.
23. Ibid., 870a.
24. Ibid.
25. Ibid.
26. Ibid., 911b.
27. Ibid., 913a.
28. Ibid., 113a.

Chapter 11: Egyptian Phonetic Values

1. Budge, *An Egyptian Hieroglyphic Dictionary,* 58b.
2. Ibid., 124a.
3. Ibid., 152a.
4. Ibid., 154b.

Chapter 12: Revisiting the Symbolism of Dogon Cosmology

1. Budge, *An Egyptian Hieroglyphic Dictionary,* 543a.
2. Ibid., 349b.
3. Ibid., 349a.
4. Ibid., 266a.
5. Genevieve Calame-Griaule, *La Parole du Monde,* 16–17.
6. Griaule and Dieterlen, *The Pale Fox,* 185.
7. Ibid., 81.
8. Budge, *An Egyptian Hieroglyphic Dictionary,* 256a.

9. Ibid., 255b.
10. Ibid., 253b.
11. Ibid., 254b.
12. Griaule and Dieterlen, *The Pale Fox,* 81–82.
13. Budge, *An Egyptian Hieroglyphic Dictionary,* 36a.
14. Ibid., 114b.
15. Ibid.
16. Helene Hagan, *The Shining Ones: A Mythological Essay on the Amazing Roots of Egyptian Civilization* (Philadelphia: Xlibris, 2000), 50.
17. Griaule and Dieterlen, *The Pale Fox,* 85.
18. Budge, *An Egyptian Hieroglyphic Dictionary,* 6a.
19. Ibid., 266a.
20. Ibid., 899b.
21. Ibid., 899b.
22. Ibid., 20ab.
23. Ibid., 899a.
24. Ibid., 896a.
25. Ibid.
26. Ibid.
27. Ibid., 897a.
28. Ibid., 910a.
29. Ibid., 908b.
30. Ibid., 910a.
31. Ibid. 909b.
32. Ibid., 909a.
33. Griaule and Dieterlen, *The Pale Fox,* 82.
34. Budge, *An Egyptian Hieroglyphic Dictionary,* 6a.
35. Ibid., 54a.
36. Scranton, *The Science of the Dogon,* 87.
37. Budge, *An Egyptian Hieroglyphic Dictionary,* 526a.
38. Ibid., 213b.
39. Ibid., cxlii.
40. Griaule and Dieterlen, *The Pale Fox,* 88.
41. Ibid., 89.
42. Budge, *An Egyptian Hieroglyphic Dictionary,* cxxx.
43. Ibid., 75a.
44. Ibid., 133a.
45. Ibid., 151a.
46. Ibid., 153a.

47. Ibid., 881b.
48. Ibid., 821b.
49. Ibid.
50. Ibid., 816a.
51. Ibid., 816a.
52. Ibid., 821b.
53. Ibid., 816b.
54. Ibid., 865a.
55. Ibid., 417b, 418a.
56. Ibid., 418b.
57. Ibid., 588a.
58. Ibid.
59. Ibid., 61a.
60. Ibid.
61. Calame-Griaule, *Dictionnaire Dogon,* 12.
62. Budge, *An Egyptian Hieroglyphic Dictionary,* 864b.
63. Ibid., cxxv.
64. Ibid.
65. Ibid., 174b.
66. Ibid., 174a, 171b.
67. Ibid., 130b.
68. Ibid., 432a.
69. Ibid., 432b.
70. Ibid., 176b.
71. Ibid., 176a.
72. Ibid., 177a.
73. Ibid.
74. Ibid., 167a.
75. Ibid., 165a.
76. Ibid., 164b.
77. Ibid., 834a.
78. Ibid.
79. Ibid., cxxxix.
80. Ibid., 706b.
81. Ibid., 713b.
82. Ibid., 707a.
83. Griaule and Dieterlen, *The Pale Fox,* 420.
84. Budge, *An Egyptian Hieroglyphic Dictionary,* 913b.
85. Ibid., 893a.

86. Ibid., cxix, 893a.
87. Ibid., 893b.
88. Ibid., 104b.
89. Ibid., 893a.
90. Ibid., 913b.
91. Ibid., 911b.
92. Ibid., 240b.
93. Ibid., 240a.
94. Ibid., 241a.
95. Ibid., 241b.
96. Ibid., 401a.
97. Ibid., cvii.
98. Ibid., 784b.
99. Ibid., 791b.
100. Ibid., 605b.
101. Ibid., 604b.
102. Ibid., 603a.
103. Ibid., 604b.
104. Ibid., cxxvi.
105. Scranton, *The Science of the Dogon,* 94.
106. Budge, *An Egyptian Hieroglyphic Dictionary,* 34b.
107. Ibid., 604a.
108. Ibid., 605b.
109. Griaule and Dieterlen, *The Pale Fox,* 130–31.
110. Budge, *An Egyptian Hieroglyphic Dictionary,* 231a.
111. Ibid., 230b.
112. Ibid.
113. Ibid.
114. Ibid., 233b.
115. Ibid.

Chapter 13: Conclusion

1. Budge, *An Egyptian Hieroglyphic Dictionary,* 910a.
2. Ibid., 401a.
3. Ibid., 408b.
4. Ibid., 848a.
5. Ibid., 408a.
6. Ibid., 546a.

7. Ibid., 547b.
8. Ibid., 548a.
9. Ibid., 546a.
10. Calame-Griaule, *Dictionnaire Dogon*, 153.
11. Budge, *An Egyptian Hieroglyphic Dictionary*, 240b.
12. Ibid., 240a.
13. Helene Hagan, *The Shining Ones: An Etymological Essay on the Amazigh Roots of Egyptian Civilization*, 48.
14. Ibid., 25.
15. Ibid., 15.
16. Ibid., 50.
17. Budge, *An Egyptian Hieroglyphic Dictionary*, 655a.
18. Ibid.
19. Ibid., 655b.
20. Ibid., cxxv.

BIBLIOGRAPHY

Allen, James P. *Middle Egyptian: An Introduction to the Language and Culture of Hieroglyphs*. Cambridge: Cambridge University Press, 2000.

Budge, E. A. Wallis. *An Egyptian Hieroglyphic Dictionary*. New York: Dover Publications, 1978.

Calame-Griaule, Genevieve. *Dictionnaire Dogon*. Paris: Librarie C. Klincksieck, 1968.

———. *La Parole du monde*. Paris: Mercure de France, 2002.

———. *Words and the Dogon World*. Philadelphia: Institute for the Study of Human Issues, 1986.

Dieterlen, Germaine. *Les Dogon: Notion de Personne et Mythe de la Creation*. Paris: L'Harmattan, 1999.

Erman, Adolf. *Altagyphtisches Worterbuch*. Leipzig: n.p., 1907.

Greene, Brian. *The Elegant Universe*. New York: Vintage Books, 2000.

Griaule, Marcel. *Conversations with Ogotemmeli*. Oxford: Oxford University Press, 1970.

Griaule, Marcel, and Germaine Dieterlen. *The Pale Fox*. Chino Valley, Ariz.: Continuum Foundation, 1986.

Hagan, Helene. *The Shining Ones: An Etymological Essay on the Amazigh Roots of Egyptian Civilization*. Philadelphia: Xlibris, 2000.

Hawking, Stephen W. *A Brief History of Time*. Toronto: Bantam, 1988.

Scranton, Laird. *The Science of the Dogon: Decoding the African Mystery Tradition*. Rochester, Vt.: Inner Traditions, 2006.

Snodgrass, Adrian. *The Symbolism of the Stupa*. Delhi: Motilal Banarsidass Publishers, 1992.

INDEX

BOOKS OF RELATED INTEREST

The Science of the Dogon
Decoding the African Mystery Tradition
by Laird Scranton
Foreword by John Anthony West

The Sirius Mystery
New Scientific Evidence of Alien Contact 5,000 Years Ago
by Robert Temple

Shamanic Wisdom in the Pyramid Texts
The Mystical Tradition of Ancient Egypt
by Jeremy Naydler

Before the Pharaohs
Egypt's Mysterious Prehistory
by Edward F. Malkowski

The Spiritual Technology of Ancient Egypt
Sacred Science and the Mystery of Consciousness
by Edward F. Malkowski

Sacred Science
The King of Pharaonic Theocracy
by R. A. Schwaller de Lubicz

The Egyptian Miracle
An Introduction to the Wisdom of the Temple
by R. A. Schwaller de Lubicz
Illustrated by Lucie Lamy

Forbidden History
Prehistoric Technologies, Extraterrestrial Intervention, and the Suppressed Origins of Civilization
Edited by J. Douglas Kenyon